TRAINING MANUAL ON FOOD IRRADIATION TECHNOLOGY AND TECHNIQUES

Second Edition

The following States are Members of the International Atomic Energy Agency:

AFGHANISTAN
ALBANIA
ALGERIA
ARGENTINA
AUSTRALIA
AUSTRIA
BANGLADESH
BELGIUM
BOLIVIA
BRAZIL
BULGARIA
BURMA
BYELORUSSIAN SOVIET
   SOCIALIST REPUBLIC
CANADA
CHILE
COLOMBIA
COSTA RICA
CUBA
CYPRUS
CZECHOSLOVAKIA
DEMOCRATIC KAMPUCHEA
DEMOCRATIC PEOPLE'S
   REPUBLIC OF KOREA
DENMARK
DOMINICAN REPUBLIC
ECUADOR
EGYPT
EL SALVADOR
ETHIOPIA
FINLAND
FRANCE
GABON
GERMAN DEMOCRATIC REPUBLIC
GERMANY, FEDERAL REPUBLIC OF
GHANA
GREECE
GUATEMALA
HAITI

HOLY SEE
HUNGARY
ICELAND
INDIA
INDONESIA
IRAN
IRAQ
IRELAND
ISRAEL
ITALY
IVORY COAST
JAMAICA
JAPAN
JORDAN
KENYA
KOREA, REPUBLIC OF
KUWAIT
LEBANON
LIBERIA
LIBYAN ARAB JAMAHIRIYA
LIECHTENSTEIN
LUXEMBOURG
MADAGASCAR
MALAYSIA
MALI
MAURITIUS
MEXICO
MONACO
MONGOLIA
MOROCCO
NETHERLANDS
NEW ZEALAND
NICARAGUA
NIGER
NIGERIA
NORWAY
PAKISTAN
PANAMA
PARAGUAY
PERU

PHILIPPINES
POLAND
PORTUGAL
QATAR
ROMANIA
SAUDI ARABIA
SENEGAL
SIERRA LEONE
SINGAPORE
SOUTH AFRICA
SPAIN
SRI LANKA
SUDAN
SWEDEN
SWITZERLAND
SYRIAN ARAB REPUBLIC
THAILAND
TUNISIA
TURKEY
UGANDA
UKRAINIAN SOVIET SOCIALIST
   REPUBLIC
UNION OF SOVIET SOCIALIST
   REPUBLICS
UNITED ARAB EMIRATES
UNITED KINGDOM OF GREAT
   BRITAIN AND NORTHERN
   IRELAND
UNITED REPUBLIC OF
   CAMEROON
UNITED REPUBLIC OF
   TANZANIA
UNITED STATES OF AMERICA
URUGUAY
VENEZUELA
VIET NAM
YUGOSLAVIA
ZAIRE
ZAMBIA

The Agency's Statute was approved on 23 October 1956 by the Conference on the Statute of the IAEA held at United Nations Headquarters, New York; it entered into force on 29 July 1957. The Headquarters of the Agency are situated in Vienna. Its principal objective is "to accelerate and enlarge the contribution of atomic energy to peace, health and prosperity throughout the world".

© IAEA, 1982

Printed by the IAEA in Austria
August 1982

TECHNICAL REPORTS SERIES No. 114

TRAINING MANUAL ON FOOD IRRADIATION TECHNOLOGY AND TECHNIQUES

Second Edition

PREPARED BY THE
JOINT FAO/IAEA DIVISION OF ISOTOPE
AND RADIATION APPLICATIONS OF
ATOMIC ENERGY FOR
FOOD AND AGRICULTURAL DEVELOPMENT

INTERNATIONAL ATOMIC ENERGY AGENCY
VIENNA, 1982

TRAINING MANUAL ON FOOD IRRADIATION TECHNOLOGY
AND TECHNIQUES, SECOND EDITION
IAEA, VIENNA, 1982
STI/DOC/10/114/2
ISBN 92–0–115082–2

FOREWORD

The food irradiation process has been under development since the mid-1940s. It is basically the newest food process since the invention of canning by Nicholas Appert over 150 years ago and has already produced many useful results. When fully applied, food irradiation will be an important means of preventing food spoilage and losses and will improve food distribution. The technique will also improve the hygienic and nutritional standards of foods.

The present publication is a revision of the Training Manual on Food Irradiation Technology and Techniques published by the IAEA in 1970. The revision retains the basic organization and scope of the first edition; it now also reflects the important advances made in the technology of food irradiation, in the radiation chemistry of foods, in the microbiology of irradiated foods and in wholesomeness and standardization. The newer concept of treating irradiation as a general food process is presented.

The practical part of the Manual, which contains a series of detailed laboratory exercises in the use of ionizing radiation for food processing, has been revised and expanded. The exercises mainly represent the efforts of the instructors who participated in the First and Second Training Courses in Food Irradiation, held by the International Facility for Food Irradiation Technology (IFFIT) in 1979 and 1980, respectively, at Wageningen, Netherlands.

The sponsoring organizations hope that the contents of this revised Training Manual will help scientists in acquiring the necessary knowledge for performing proper research and development work in the field of food irradiation. The Manual presents an up-to-date picture of the current state of food irradiation and is thus very useful to everyone interested in the practical application of food processing by irradiation.

ACKNOWLEDGEMENTS

A draft manuscript of the revised Manual was prepared by Walter Urbain, Professor Emeritus, Michigan State University, USA.

Valuable contributions to some parts of the Manual were rendered by the following:

K.H. Chadwick, Foundation Institute for Atomic Science in Agriculture, Wageningen, Netherlands

J.F. Diehl, Bundesforschungsanstalt für Ernährung, Karlsruhe, Federal Republic of Germany

P. Elias, International Project in the Field of Food Irradiation, Karlsruhe, Federal Republic of Germany

J. Farkas, International Facility for Food Irradiation Technology, Wageningen, Netherlands

H.P. Leenhouts and J.F. Stoutjesdijk, Foundation Institute for Atomic Science in Agriculture, Wageningen, Netherlands

The manuscript in its final form was prepared by J.G. van Kooij of the Joint FAO/IAEA Division of Isotope and Radiation Applications of Atomic Energy for Food and Agricultural Development, IAEA, Vienna. Substantial assistance was obtained from P. Loaharanu, of the same Division.

CONTENTS

PART I – LECTURE MATTER

PART II – LABORATORY EXERCISES

PART I

LECTURE MATTER

LECTURE MATTER

1. RADIONUCLIDES AND RADIATION

Let us start the study of the ionizing radiation phenomenon that is the basis of the food irradiation process by discussing the atomic model.

1.1. Atomic model: definitions

An atom is composed of a positively charged nucleus which is surrounded by shells of negatively charged (orbital) electrons. The nucleus contains *protons* and *neutrons* as its major components of mass; the former have a positive charge, and the latter have no charge. The nucleus has a diameter of approximately 10^{-14} m and contains almost the entire mass of the atom. The atom, including the orbital electrons, has a diameter of approximately 10^{-10} m.

The number of protons (Z) in the nucleus, which is characteristic of a chemical element, is called the atomic number (proton number). The atoms of a particular element may not, however, all have the same number of neutrons (N) in the nucleus. Atom types that have the same Z-values but different N-values are called isotopes of the same element. They occupy the same place in the periodic chart of the elements. As the neutrons and protons represent the major part of the mass of the atom and each has an atomic weight, i.e. an atomic mass close to unity, the mass number, which is the sum of protons and neutrons, is close to the atomic mass **M**:

$$\text{Mass number} = Z + N \doteq M$$

The nuclei of some isotopes are not stable; they disintegrate spontaneously at a characteristic decay rate. In nature, a number of unstable isotopes are known and nowadays many unstable isotopes are produced artificially in atomic reactors and by particle accelerators. As the disintegration of unstable isotopes is accompanied by the emission of various kinds of radiation, these unstable isotopes are called radionuclides or radioisotopes.

The nuclei of radionuclides may emit α-, β^{+}-, β^{-}- and γ-rays. The α-particles are fast-moving He nuclei, each containing two protons and two neutrons. β^{+}- and β^{-}-particles are, respectively, positively and negatively charged high-speed electrons, while γ-rays are electromagnetic wave packets (photons) of very short wavelength compared with visible light, but travelling at the speed of light.

Natural isotopes of low Z (except ordinary hydrogen) have approximately the same number of neutrons as protons ($N \doteq Z$) in their nuclei, and they are usually stable. As the atomic number of the elements increases, the number of neutrons increasingly exceeds the number of protons, which finally results in unstable nuclei. Thus, the majority of unstable isotopes in nature are found for elements of high

Z-number with a neutron:proton ratio of the order of $1\frac{1}{2}$:1. The emission of
α-particles is characteristic of these elements. The combination of two protons and
two neutrons is one of the very stable nuclear forms; and this combined form, the
α-particle, is ejected as a single particle from the nucleus of the radioactive atom.

There appears to be a more or less well defined optimum N:Z ratio for the
stability of each element. When the number of neutrons in the nucleus of a radionuclide is excessive, the number of protons in the nucleus tends to increase by the
ejection of a negative β-particle (negatron) and a neutrino from the nucleus.
This β-particle accompanies the transformation of a neutron into a proton:

$$n \rightarrow p^+ + \beta^- \quad (+ \text{ neutrino})$$

An excess of protons in a nucleus may be counteracted by the ejection of a *positron,*
a positively charged electron (regarding MeV, see Section 1.3):

$$p^+ + 1.02 \text{ MeV} \rightarrow n + \beta^+ \quad (+ \text{ neutrino})$$

An excess of protons in the nucleus may, alternatively, be reduced by the
nucleus capturing one of its own orbital electrons, a process known as electron
capture (EC) or K-capture since the electron is captured from the innermost or
K-shell of orbital electrons:

$$p^+ + e^- \rightarrow n + \text{neutrino}$$

EC is accompanied by the emission of a characteristic X-ray, most frequently
representing the energy difference between an L- and a K-shell electron in the
element formed (a 'hole' in the K-shell being filled by an L-electron).

After the ejection of an α- or β-particle, or after EC, the energy level of the
daughter nucleus may not be at its ground state. The excess energy of this excited
nucleus is emitted in the form of one or more γ-photons.

The excited nucleus may interact with an orbital electron in the decaying atom,
whereby the electron is ejected from the atom at a given velocity, and the expected
γ-photon is not emitted. This process results in the combined emission of a fast
electron and a characteristic X-ray and is known as internal conversion (IC).
The X-ray photon may in turn undergo IC, producing what is called an Auger
electron. When a large nucleus such as ^{235}U captures a neutron, the nucleus will
divide into two parts of different masses. This process is called fission and is
accompanied by the release of neutrons.

All the primary fission products are unstable (excessive N), and each forms a
series of radioactive daughter nuclides terminating with a naturally occurring stable
isotope.

Summarizing, we may say that radionuclides will emit particles and/or photons
of the following nature:

α-particle	— doubly positively charged particle, containing two neutrons and two protons and originating at high speed from the nucleus;
β⁻-particle	— high-speed electron from the nucleus, negatively charged;
β⁺-particle	— high-speed positron from the nucleus, positively charged;
γ-ray photon	— electromagnetic energy packet coming from the nucleus at the speed of light;
X-ray photon	— electromagnetic energy packet coming from an electron shell at the speed of light, following EC or IC;
Auger electron	— electron resulting from an X-ray photon upon internal conversion;
IC electron	— (internal conversion electron) electron emitted as a result of the interaction between an excited nucleus and an electron of the inner shells;
Neutron	— particle with no charge and a mass close to that of a proton.

1.2. Radioactive decay and 'specific activity'

The number of disintegrations per unit increment of time is a constant fraction of the number of radioactive atoms present at that time. Mathematically this can be expressed as

$$D^* = -\frac{dN^*}{dt} = \lambda^* N^* \qquad (1)$$

where

D^* is the disintegration rate (expressed per unit time) at time t;
N^* is the number of radioactive atoms present at time t; and
λ^* is the decay constant expressed in reciprocal time units.

The minus sign indicates that the number of radioactive atoms decreases with time t. Integrating the differential equation (1) and calling the number of radioactive atoms present at beginning time N_0, one obtains

$$N^* = N_0^* e^{-\lambda^* t} \quad \text{or} \quad D^* = D_0^* e^{-\lambda^* t} \qquad (2)$$

It follows from Eq. (2) that the time required for one-half of the original activity to decay is independent of the beginning number of atoms. Designating the time required for half decrease of original activity as $t_{\frac{1}{2}}$, one obtains

$$\tfrac{1}{2} D_0^* = D_0^* e^{-\lambda^* t_{\frac{1}{2}}}; \ \text{i.e.} \ \lambda^* t_{\frac{1}{2}} = \ln 2 = 0.693$$

where $t_{\frac{1}{2}}$ is the half-life of the isotope expressed in time units. It is seen that the product of decay constant and half-life of any isotope is 0.693, which is useful for conversion of $t_{\frac{1}{2}}$ to λ^*. The decay constant, having the dimension of reciprocal time and usually being a small number, is inconvenient for many purposes. Instead, half-life ($t_{\frac{1}{2}}$ in, for example, days or years) is often used as the decay characteristic of a radionuclide.

The special unit of activity (radioactivity) has for many years been the curie (abbreviated to Ci). This was originally defined as the radioactivity associated with the quantity of radon in equilibrium with 1 gram of radium (1910). The formal definition agreed in 1964, when the curie was accepted for use with the International System of Units (SI), was:

$$1 \ \text{Ci} = 3.7 \times 10^{10} \ \text{disintegrations per second}$$
$$= 3.7 \times 10^{10} \, \text{s}^{-1} \ \text{(exactly)}$$

Since 1976, a new unit of activity, the becquerel (Bq), has been defined as a derived unit of the International System of Units (SI):

$$1 \ \text{Bq} = 1 \ \text{disintegration per second}$$
$$= 1 \, \text{s}^{-1}$$

Hence

$$1 \ \text{Ci} = 3.7 \times 10^{10} \ \text{Bq}$$
$$1 \ \text{Bq} = 2.7027 \times 10^{-11} \ \text{Ci}$$

The old special unit, the curie, is to be phased out in the next few years.

If one has g^* grams of a radioisotope with a decay constant λ^* and an atomic mass M, the radioactivity expressed in curies will be as follows (N^0 is Avogadro's number):

$$\frac{g^*}{M} \times N^0 \qquad = \text{total number of radioactive atoms } (N^*)$$

$$\lambda^* \times \frac{g^*}{M} \times N^0 \qquad = \text{total disintegrations per minute } (D^*) \text{ if } \lambda \text{ is given in min}^{-1}$$

$$\lambda^* \times \frac{g^*}{M} \times \frac{N^0}{2.22 \times 10^{12}} \qquad = \text{total activity in curies}$$

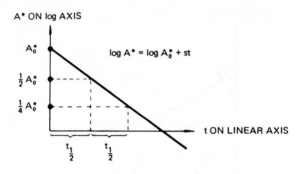

FIG.1. *Decay curve of a single radionuclide.*

The decay constant or the half-life of an isotope can be graphically determined if the half-life is within a measurable range. It appears from Eq. (2) that if the measured activity $A^* = YD^*$ (where Y is the constant counting yield) is plotted against time on semi-log paper, a straight line will be observed. The half-life or decay constant can easily be found directly (see Fig.1) or from the slope s, which is equal to $-\lambda^*/2.3$. For isotopes of very long half-life, one has to apply the method of absolute measurement for half-life determination.

When two radioisotopes, A and B, are present simultaneously, the observed activity is

$$A_0^* e^{-\lambda_A^* t} + B_0^* e^{-\lambda_B^* t}$$

If this activity is plotted on semi-log paper, one obtains a composite curve, such as appears in Fig.2. With the assumption that the half-lives are sufficiently different (e.g. a factor of 10), the curve can be resolved graphically by subtraction of the extrapolated straight line resulting from the long-lived component (B) from the sum curve observed. The two straight lines then yield the two half-lives.

In practice, a radionuclide will be accompanied by a variable quantity of stable isotopes of the same element. The stable form is called 'carrier'. To specify the concentration of radionuclide in one element or compound, the term *specific activity* is introduced. This is generally expressed as radioactivity per unit amount of specified test substance.

By some procedures, radionuclides can be prepared virtually free from carrier, in which case they are called carrier-free.

1.3. Energy of radiation

The energy unit commonly used for radiation is the electronvolt (eV). This is equivalent to the kinetic energy acquired by an electron on being accelerated

7

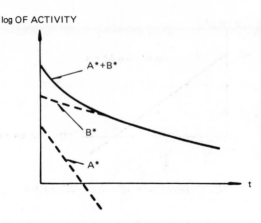

log OF ACTIVITY

A*+B*

B*

A*

t

FIG.2. Decay curves of two radionuclides, A and B, simultaneously present in a sample.

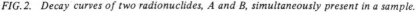

through a potential difference of one volt. 1 keV and 1 MeV are 10^3 eV and 10^6 eV, respectively; 1 MeV $\cong$ 1.6 $\times$ 10^{-13} joule $\cong$ 1.6 $\times$ 10^{-6} erg.

The kinetic and the total energies, respectively, of the particles and photons emitted by radionuclides have characteristic values, which are usually indicated on nuclear charts for each isotope. Any energy spectrum of the α-particles, γ-photons or characteristic X-ray photons emitted by a radionuclide is *discrete,* showing one or a few monoenergetic ('monochromatic') lines. On the other hand, the energy of β-particles ejected by a given nuclide varies from zero up to a certain maximum energy (E_{max}) that is at the disposal of the β-particle. This is because a variable part of E_{max} is taken away by a neutrino or an antineutrino, neither of which is observable in ordinary counting (they have no charge and practically no mass). As a consequence, the β-particles show a *continuous* spectrum of energies from zero up to the characteristic E_{max}. The β-energies given in a table or chart of nuclides are E_{max} values; the average β-particle energy is usually about one-third E_{max}. The continuous β-spectrum may sometimes be overlapped by one or two monoenergetic lines from IC electrons.

The characteristic radiations and energies for a given radioisotope are often shown in the form of decay schemes (see, e.g., Fig.3).

A knowledge of decay characteristics is important in the study of protection against and measurement of radionuclides.

1.4. Interaction of radiation with matter

1.4.1. Absorption of α-particles

The α-particles ejected from any particular radionuclide are monoenergetic. In passing through matter and interacting with the atoms thereof, the kinetic energy

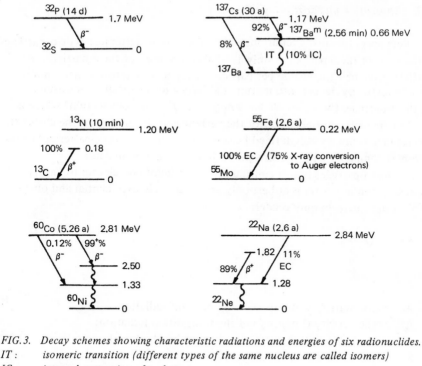

FIG. 3. Decay schemes showing characteristic radiations and energies of six radionuclides.

IT : isomeric transition (different types of the same nucleus are called isomers)

IC : internal conversion of γ-photon

EC: electron capture (K-capture)   ↧ :   γ-photon

↘ : β⁻-particle   h :   hour(s)

↙ : β⁺-particle   d :   day(s)

↙EC: electron capture   a :   year(s)

$^{137}Ba^m$:excited ^{137}Ba, called 'metastable' because the emission of the γ-ray is not instantaneous

of the α-particle will be spent in (1) exciting outer-shell electrons to higher-energy orbits, and (2) ejecting electrons out of their orbits. Since α-particles are doubly charged and the mass is relatively large (atomic mass 4), a dense track of ion pairs (i.e. ejected electrons and positively charged atom residues) is formed along the path of an α-particle. As the α-particle dissipates its energy along its path, the velocity of the particle decreases and finally the particle acquires two electrons from its surroundings and becomes a helium atom. The *range,* i.e. the distance that an α-particle can penetrate into any matter (absorber), depends on the initial energy of the particle and the density of the absorber. The range of the α-particle is generally small and amounts to several cm in air and several μm in aluminium for energies of the order of 1–10 MeV. As the energy of an α-particle is lost in a relatively thin layer of absorber, it is evident that the number of ion pairs per cm of track, the *specific ionization,* is very high.

1.4.2. Absorption and scattering of β-particles

Beta particles cause excitations and ionizations in matter just as do α-particles, but the mass of the β-particle is only 1/7000 of the mass of the α-particle, and β-particles have half the charge per particle. They will therefore scatter more, penetrate relatively deeper into matter, and have a lower specific ionization. Like the α-particle, the β-particle has a *range* (see the previous section) which is characteristic of the initial energy of the particle and the density of the absorber, but this range is not so well defined because of the zig-zag path (scattering) of the electron as compared with the straight path of the helium nucleus.

Because β-particles have a continuous spectrum of energies up to an E_{max}, their absorption in matter is at best only approximately exponential and obeys the following equation only crudely:

$$A^* = A_0^* e^{-\mu d}$$

where

A_0^* is the activity (intensity) of the incident radiation;
A^* is the activity (intensity) of the transmitted radiation;
μ is the β-absorption coefficient of the absorber; and
d is the thickness of the absorber.

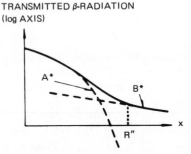

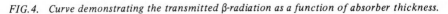

FIG.4. *Curve demonstrating the transmitted β-radiation as a function of absorber thickness.*

Therefore, when the radiation transmitted by the absorber is plotted as a function of the absorber thickness on semi-log paper, a fairly straight line is obtained over a portion of the curve (Fig.4).

The curve becomes practically horizontal at R, the range for β-particles with E_{max}. Although all the β-rays are stopped at this absorber thickness, one still finds some transmission of radiation because, particularly at low velocities, the β-particles interact with the atoms of the absorber, giving rise to (non-characteristic) X-rays, the *bremsstrahlung* (B*). By subtraction of B* from the composite curve, the pure β-transmission curve (A*) is obtained.

Positron energy absorption takes place in the same manner as for negative β-radiation. However, when the kinetic energy of the positron becomes very low, the positron is *annihilated* together with an electron, giving rise to two characteristic photons of 0.51 MeV each: $e^+ + e^- \rightarrow 2$ photons.

Absorption and scattering of β-particles is important in the measurement of β-active samples. Absorption and scattering will occur in a sample cover or a detector window as well as in intervening air. Side-scattering (into the detector) from a counter shield and/or back-scattering from a sample support will also occur. These effects will all influence the counting rate one way or another. Finally, unless the sample is 'infinitely' thin, self-scattering (into and away from the detector) and self-absorption will all take place in the material of the sample itself, and this will cause an overall *self-weakening* effect, which is largest for thick samples and small (even slightly negative) for very thin samples. The counting rate from samples of increasing thickness at first increases because of greater total activity and then becomes constant (at 'infinite' thickness) because the contribution of β-activity from the lower layers of the sample is entirely absorbed in the upper ones.

1.4.3. Attenuation of γ- and X-rays

In passing through matter, the energy of γ- and X-ray photons is attenuated by three important interactions: (1) photoelectric effect, (2) Compton scattering and (3) pair-production.

(1) When the photon energy is below about 0.5 MeV, the photoelectric effect is predominant. The total energy (i.e. the entire photon) is used up in the ejection of an electron at high speed from an atom shell. Subsequently, this fast electron causes many excitations and ionizations just as a β-particle does. The photoelectric effect is particularly important when the atoms of the absorber have a high Z-number.

(2) Compton scattering arises predominantly when γ-photons in the energy range 0.5–5 MeV collide with free or loosely bound electrons in the absorber. Part of the photon energy is transferred to the electron as kinetic energy in such a collision, and the reduced photon is deflected (slightly or up to 180°) from its original direction. This effect is important for absorber atoms of high Z-number.

(3) When a photon has an energy of at least 1.02 MeV or higher, it may become extinct in the proximity of an atomic nucleus of the absorber, giving rise to an electron-positron pair. Any photon energy above the required 1.02 MeV is imparted to the e^- and the e^+ as kinetic energy.

Theoretically, γ- or X-radiation is never completely stopped by matter, although the transmitted radiation may be reduced to an insignificant value. For

11

a collimated beam of monoenergetic photons, attenuation by absorption and scattering can be described mathematically as follows:

$$I = I_0 e^{-\mu d}$$

where

I_0 is the initial intensity of collimated monoenergetic photons;
I is the intensity after passing d cm of the absorber; and
μ is the attenuation coefficient for the photon energy and the material concerned. (If d is expressed in cm, μ should be given in cm^{-1}.)

This is the well-known Lambert-Beer law for visible light photons. The derivation of the equation from the basic assumption that $-dI/dx = \mu I$ is analogous to the derivation of the radioactive decay law $N^* = N_0^* e^{-\lambda^* t}$ (see Section 1.2).

The numerical value of the linear attenuation coefficient μ is dependent on the γ-photon energy and the type of absorber material. That is why the mass attenuation coefficient μ/ρ is often used, expressed in $cm^2 \cdot g^{-1}$ (ρ is the density of the absorbing material in $g \cdot cm^{-3}$). The advantage of using this mass attenuation coefficient is that μ/ρ is approximately constant for all materials for γ-energies between 1 MeV and 3 MeV. When using μ/ρ, the absorber thickness should be expressed in $g \cdot cm^{-2}$, which is found by multiplying the thickness of the absorbing layer in cm by the density in $g \cdot cm^{-3}$.

From $I = I_0 e^{-\mu d}$ a half thickness value or half value layer, $d^{\frac{1}{2}}$ can be derived. $d^{\frac{1}{2}}$ is the thickness of the layer of absorbing material at which I is reduced to one-half of its initial value:

$$d^{\frac{1}{2}} = \frac{\ln 2}{\mu} = \frac{0.693}{\mu}$$

The same relation exists for μ/ρ.

The mass attenuation coefficients for water, concrete and lead for ^{60}Co (γ:1.17 MeV and 1.33 MeV) and ^{137}Cs (γ:0.662 MeV) are given in Table I. The equation $I = I_0 e^{-\mu d}$, describing the exponential attenuation, holds for narrow-beam geometry.

When the absorbing material is extensive in all directions, the scattering of radiation will be greater. This can be expressed by the 'build-up factor' B. This factor is a function of the material itself, the thickness of the material, photon energy and the geometry of the radiation field. Thus, for broad-beam conditions the attenuation can be described by

$$I(d) = BI_0 e^{-\mu d}$$

TABLE I. MASS ATTENUATION COEFFICIENTS μ/ρ (in $cm^2 \cdot g^{-1}$)

Radionuclide	Water	Concrete	Lead
Co-60	0.064	0.049	0.061
Cs-137	0.084	0.078	0.11

Data from Radiological Health Handbook, US Dept. of Health, Education and Welfare, Public Health Service, revised edn (1970) p. 139.

B is dependent on μd and on the absorber material (see Table II). In Figs 5 and 6 the transmission is given as a function of the absorber thickness. An understanding of the interactions of high-energy electromagnetic radiation with matter is necessary where shielding, dose calculations and measurement of γ-, X- and annihilation photons are concerned.

1.4.4. Scattering and absorption of neutrons

Neutrons, being without charge, lose energy only by direct contact with nuclei of matter. The processes may be of the following four types:

(1) Of an elastic nature, like billiard-ball collisions. Ion pairs are produced by these collisions, the hit nucleus losing one or more of its orbital electrons. Neutrons of high initial energy (fast neutrons) gradually lose their energy by this interaction until they have been moderated to 'slow' or 'thermal' neutrons. Light elements, especially H, have the best neutron-moderating qualities.

(2) Of a type in which the neutron is absorbed by nuclei with resultant nuclear reaction. This occurs predominantly with slow neutrons, e.g.

$$^{10}B + n \rightarrow (^{11}B) \rightarrow {}^7Li + \alpha + \gamma$$

(3) When the nuclei of certain elements of high atomic number are hit by neutrons of appropriate energy, fission results (the nuclear pile).

(4) Finally, free neutrons decay spontaneously, with a half-life of 12 minutes, to protons and β-particles, which thereupon excite and ionize atoms of matter.

1.4.5. Induced radioactivity

When ionizing radiation impinges on matter, energy may be imparted to the nuclei of some of the atoms. Under certain conditions, this may cause excitation

13

TABLE II. EXPOSURE BUILD-UP FACTOR FOR ISOTROPIC POINT SOURCES

Material	Photon energy, E (MeV)	μd						
		1	2	4	7	10	15	20
Water	0.255	3.09	7.14	23.0	72.9	166	456	982
	0.5	2.52	5.14	14.3	38.3	77.6	178	334
	1.0	2.13	3.71	7.68	16.2	27.1	50.4	82.2
	2.0	1.83	2.77	4.88	8.46	12.4	19.5	27.7
	3.0	1.69	2.42	3.91	6.23	8.63	12.8	17.0
Aluminium	0.5	2.37	4.24	9.47	21.5	38.9	80.8	141
	1.0	2.02	3.31	6.57	13.1	21.2	37.9	58.5
	2.0	1.75	2.61	4.62	8.05	11.9	18.7	26.3
	3.0	1.64	2.32	3.78	6.14	8.65	13.0	17.7
Iron	0.5	1.98	3.09	5.98	11.7	19.2	35.4	55.6
	1.0	1.87	2.89	5.39	10.2	16.2	28.3	42.7
	2.0	1.76	2.43	4.13	7.25	10.8	17.6	25.1
	3.0	1.55	2.15	3.51	5.85	8.51	13.5	19.1
Lead	0.5	1.24	1.42	1.69	2.00	2.27	2.65	2.73
	1.0	1.37	1.69	2.26	3.02	3.74	4.81	5.86
	2.0	1.39	1.76	2.51	3.66	4.84	6.87	9.00
	3.0	1.34	1.68	2.43	3.75	5.30	8.44	12.3

Data from Radiological Health Handbook, US Dept. of Health, Education and Welfare, Public Health Service, revised edn (1970) pp. 145, 146.

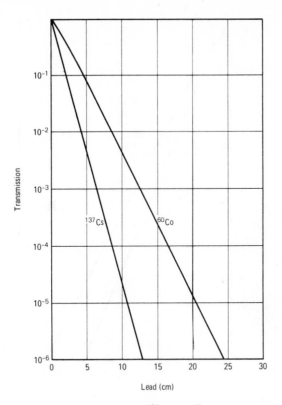

FIG. 5. Broad-beam transmission of γ-rays from ^{137}Cs and ^{60}Co through lead, density 11.35 g·cm^{-3}. Data derived from Radiological Health Handbook, US Dept. of Health, Education and Welfare, Public Health Service, revised edn (1970) p.148.

sufficient to induce an atomic nucleus to become so unstable that it emits a neutron, together with γ-radiation. This reaction changes the atomic nucleus into that of a different element or that of an isotope of the original one. In this way, ionizing radiation may induce the appearance of radioactivity in matter which previously showed virtually none. The possibilities of producing such induced radioactivity depend on the properties of the matter irradiated and on the energy of the radiation employed. If the energy of the radiation source is sufficiently high, several of the elements contained in food can be made radioactive. For example, ordinary ^{16}O can be induced to undergo such a change when irradiated at an energy level of 15.5 MeV, and ordinary ^{12}C can be induced to make such a change at an energy of 18.8 MeV.

Since these are among the principal elements contained in food, it is important that they do not become radioactive. It is for this reason that particular radiation sources have been selected for the irradiation of foods. At present, the radiation

15

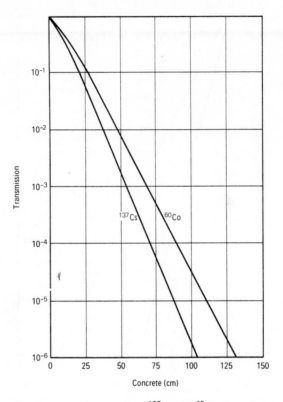

FIG.6. *Broad-beam transmission of γ-rays from* ^{137}*Cs and* ^{60}*Co through concrete, density* *2.35 g·cm*$^{-3}$. *Data derived from Radiological Health Handbook, US Dept. of Health, Education and Welfare, Public Health Service, revised edn (1970) p.149.*

sources permitted are: (1) ^{60}Co; (2) ^{137}Cs; (3) accelerated electrons of an energy not more than 10 MeV; and (4) X-rays from a source with a beam energy of not more than 5 MeV. It has been amply demonstrated that there is no danger of inducing radioactivity in food with such selected sources.

2. RADIATION DETECTION AND MEASUREMENT

Ionizing radiation interacts with all matter (gaseous, liquid or solid), causing chemical changes, ionization and excitations. These effects are utilized in the various methods of detection and measurement.

In radiography, for example, ionizing radiations are detected by their effect on photographic emulsions. In the ionization chamber, the gas-flow detector,

16

the Geiger-Müller tube and the neutron detector, and ions produced directly or indirectly by the radiation are collected on charged electrodes. In solid and liquid scintillation counting, emission photons (in the blue ultra-violet region) form the basis of detection. The phenomenon of lyoluminescence, in which previously irradiated solid phosphors emit light on dissolution, provides a method for measuring radiation. Certain chemical reactions produced by ionizing radiation can be used to measure the amount of radiation. Since absorbed ionizing radiation degrades to heat, calorimetry can be employed for quantitative measurement of radiation.

Several detection and measurement systems will now be described.

2.1. Detection by ionization

A number of detectors are based on the principle that, in an electric field, negative particles will move to a positive electrode and positive particles to a negative electrode. Charged particles that arrive at an electrode will give rise to an electronic pulse, which can be amplified and registered. Alternatively, the pulses may be merged to form an electric current, which again can be amplified and measured.

Alpha and beta particles and IC electrons (e) have a high specific ionization, i.e. produce a great number of ion pairs per unit length of track. Gamma- and X-rays have a much lower primary specific ionization, but at least one fast electron will be released by each photoelectric effect or Compton scattering (or pair production if the energy is very high), and these fast electrons will ionize just as do β-particles. Neutrons may also produce ions, directly (collision) or indirectly (following nuclear absorption), as described in Section 1.4.4. Detection by ionization of these kinds of radiation is based on the fact that atoms of a gas (in the detector) will become ionized when they are hit by the radiation particles or photons. The number of ionizations in the gas is a direct measure of the quantity of ionizing particles or photons (α, β, e, γ, X or n) that reach the detector. When an electric field is created in the detector, the negative ions (electrons) will start moving and, by hitting the positive electrode (anode), discharge. In the same way, the positive ions will move towards the cathode.

Four different types of ionization instrument will now be described.

2.1.1. Electroscope

In the electroscope or simple electrometer (see Fig. 7) the positive electrode is a rod with a wing or a metal string, and the negative electrode is the wall of the detector.

When the electroscope is fully charged, the deflection of the wing or string will be maximal (A), the amount of deflection being a function of the charge accumulated. When a radioactive source is brought near the detector, the air in the detector will become ionized and electrons will move in the direction from wall to rod. As a consequence, the deflection will decrease (B).

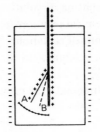

FIG. 7. Electroscope.

This type of detector is commonly used as a 'pocket dose meter' and gives a measure of the accumulated dose of external radiation (γ-, X- and hard-β-radiation) to which a worker has been exposed during a certain period.

2.1.2. Ionization chamber

Not all the ions will discharge on the electrodes of an electroscope. A certain number will recombine before they have reached the electrodes. If the voltage applied to the electrodes is steadily increased, the losses resulting from recombination will decrease, and eventually all the ions will discharge on the electrodes of the detector. If the voltage difference between the electrodes is further increased up to a certain limit, the number of ion pairs that discharge will remain constant. Each ionizing particle or photon will thus give rise to an electric pulse on the electrodes. A radiation intensity (i.e. a constant stream of particles or photons) gives rise to a continuous series of pulses, which, if allowed to merge, form a weak electric current, that may be amplified and registered by an electronic circuit. The final scale reading will then be a measure of the energy dissipated in the ionization chamber per unit of time by the ionizing particles or photons. This kind of detection instrument is thus a *dose-rate meter* (e.g. the 'cutie pie').

A small electrically charged ionization chamber, held in place for instance by a finger ring, may be used to measure accumulated exposure dose. An electronic vacuum-tube voltmeter is often necessary to measure the charge reduction, which is proportional to dose.

2.1.3. Proportional counter

If the voltage difference between the anode and the wall of the counter is increased above a certain limit, another phenomenon, known as secondary

18

ionization, will become important. The electrons that have arisen from primary ionization will produce secondary ion pairs of the gas atoms in the counter tube as they are accelerated towards the anode. This process of secondary ionization becomes increasingly important as the voltage difference between the electrodes is further increased. The final pulse size will be proportional to the energy of the initial ionizing particle (as long as all this energy is dissipated in the detector), provided the applied voltage remains constant during the measurement. Usually the radioactive sample will be placed inside the detector, which will be transfused by a gas at atmospheric pressure (*gas-flow counters*). In this way particles of low energy, such as the β^- from ^{14}C, may be counted effectively ('window-less' counting), provided suitable amplification precedes the register.

2.1.4. Geiger-Müller (G-M) counter

When the voltage difference between the electrodes of the detector is still further increased, secondary ionization becomes predominant and each primary ionizing event results in a discharge of a great number of electrons (avalanche). At this stage the large output pulse is independent of the energy of the initial particle or photon, and a further increase of the high voltage does not appreciably alter pulse size or count rate. Geiger-Müller counter detectors (G-M tubes) operate at this voltage 'plateau'. The discharges of secondary electrons initiated by one ionizing particle or photon would continue if the detector were of an open design, as in the gas-flow counter (atmospheric pressure). G-M tubes operate at a reduced gas pressure (about one-tenth atmosphere), containing a certain amount of 'quenching' gas. Usually the closure of a G-M tube is a very thin mica window $(1-3 \ mg \cdot cm^{-2})$, and the filling gas is often a noble gas like argon with, for example, alcohol or halogen as the quenching gas. A certain number of molecules are dissociated during the quenching of each discharge with alcohol. Therefore, the quantity of quenching gas in the G-M tube decreases steadily and this effect limits the life of the tube. This disadvantage does not exist when a halogen gas, e.g. chlorine, is used for quenching, because the atoms of the dissociated chlorine molecule recombine, and the life of the tube is therefore determined by other effects such as corrosion and leakage.

Energetic β- or e-particles and γ- or X-photons emitted by radioactive liquids may be counted with a thin glass wall 'dip-counter' G-M tube which is immersed in the liquid or with a specially designed liquid detector consisting of a cylindrical glass container round the G-M tube. The radioactive liquid thus surrounds the G-M tube in both cases. Particles of low energy can obviously not be counted in this way because of absorption in the wall of the G-M tube.

The fact that some time is required for each discharge of electrons (100 to 300 μs) implies that during this time no other particle or photon can be detected by the G-M tube. This time is called the *dead time* of the G-M counter, and a correction for this dead time must be made, particularly for high count rates.

Let $\mathcal{R}$ be the observed count rate and τ the dead time of the counter, in minutes. During one minute the counter will have been ineffective for $\mathcal{R}\tau$ minutes. $\mathcal{R}$ counts have therefore been registered in $1-\mathcal{R}\tau$ minutes. The corrected count rate $\mathcal{R}^+$ in cpm will therefore be $= \mathcal{R}/(1-\mathcal{R}\tau)$. When the dead time of the counter tube is known, the correction for high count rates can then be made with the aid of the above expression for $\mathcal{R}^+$. This expression is approximate, however, and should not be used to give corrections above 10%, when it is better to dilute or count at a distance from the detector.

Sometimes the dead time of a G-M tube will be fixed electronically at 300 or 400 μs so that a correction table can be used. Correction is normally not necessary unless the count rate exceeds about 2000 cpm.

$$
\begin{aligned}
\text{Numerical example:} \qquad \tau &= 300\ \mu s \\
&= 5\ \mu min \qquad \text{corr.} = 2\tfrac{1}{2}\% \\
\mathcal{R} &= 5000\ \text{cpm} \\
\mathcal{R}^+ &= 5125\ \text{cpm}
\end{aligned}
$$

G-M counters are used most widely for the detection and measurement of β-particles. For γ-rays they are not very effective (1–3% efficiency) because most of the photons will penetrate the gas without any interaction. For the detection of β-particles on glassware, benches or trays, *monitors* are used. A monitor consists of a G-M tube connected to a power unit and a count-rate meter. Often a small loud-speaker is connected to the rate meter so that a noise will warn the operator when the tube is in the vicinity of a contaminated spot.

Normally, for the assaying of activity in samples, the G-M tube will be connected to a voltage source, an amplifier, a register and a timing unit.

2.2. Detection by excitation

2.2.1. Solid scintillation counting

Solid scintillators are particularly suited for the detection of γ-rays and X-rays because of the high stopping power of the solid. Their operation is based on the following principle:

When a γ-photon interacts with a crystal, e.g. of thallium-activated NaI, at least one fast electron is liberated (see Section 1.4.3), and a constant fraction of the electron's kinetic energy is spent on excitation of orbital electrons in atoms of the crystal. On de-excitation these give rise to emission of a light-flash consisting of a number of photons. The number of light photons will be proportional to the energy dissipated in the crystal by the γ-photon.

The light photons reach the *photocathode* of a *photomultiplier*, where photoelectrons are released. The number of photoelectrons, being a constant fraction of the number of light photons, is therefore proportional to the energy originally dissipated by the γ-photon. The photocathode is connected with a series of *dynodes,* i.e. positive electrodes of increasing potential. When a photo-electron hits a dynode, secondary electrons are produced which will, in turn, hit the next dynode. In this way, the photomultiplier will, all in all, produce a large number of electrons (a pulse), proportional to the energy originally dissipated by the γ-photon in the crystal. This final pulse will be amplified linearly and registered.

As opposed to a G-M tube, the scintillation tube thus provides an output pulse that is proportional to the input energy. The scintillation tube is therefore a suitable detector for γ-ray spectrometry. A further advantage of the scintillation counter is its small dead time of only a few μs. This enables high count rates to be determined (up to at least 100 000 cpm) without the necessity to apply a correction for dead time.

For measuring β-particles, special plastic scintillators have been devised (as well as anthracene and naphthalene) which have a much higher efficiency than NaI crystals. An effective scintillator for α-particles is a thin layer of silver-activated ZnS.

2.2.2. *Liquid scintillation counting*

For counting very-low-energy and low-energy β-particles such as ^3H (0.018 MeV) and ^{14}C (0.155 MeV), a method of detection called liquid scintillation counting is often employed. For this technique, the sample to be counted is placed in solution with the scintillator so that each radioactive atom or molecule is surrounded by molecules of the scintillator. By this method, absorption is reduced and hence counting yield increases.

The scintillator system contains a solvent which is usually an organic compound, such as toluene or dioxane, and a solute which is the actual scintillator. The solvent absorbs the energy and transfers it to the solute, which then emits the light-flash. Often a secondary solute which acts as a wavelength shifter is added; i.e. it increases the wavelength of the emitted light-flash to one for which the photo-multiplier tube is more sensitive, thus increasing the counting yield.

In practice, two photomultiplier tubes facing each other across the counting chamber are often used. A coincidence circuit is employed, and only those events witnessed by both tubes are counted. This increases the signal-to-noise ratio.

Variable discriminators can be applied to this system and, since the pulse height is proportional to the input energy, pulse-height analysis is possible.

2.3. Chemical dose meters

Reactions employed in chemical dosimetry include: (1) oxidation of ferrous salts to ferric; (2) reduction of ceric salts to cerous; (3) gas evolution from aqueous solutions (of iodides); (4) acid production in chlorinated hydrocarbons; (5) the coloration of leuko dye solutions; and (6) changes in absorption, luminescence and other properties of solids such as glasses, plastics, sugars, etc.

Of these chemical dose meters the one having some degree of general acceptance is the Fricke dose meter. This involves the radiation-induced oxidation of ferrous ions in an air-saturated 0.4M sulphuric acid solution to ferric ions. The oxidation is accomplished according to the following equations:

$$Radiation + H_2O = H + OH$$

$$Fe^{2+} + OH = Fe^{3+} + OH^-$$

$$H + O_2 = HO_2$$

$$Fe^{2+} + HO_2 = Fe^{3+} + HO_2^-$$

$$HO_2^- + H^+ = H_2O_2$$

$$Fe^{2+} + H_2O_2 = OH + OH^- + Fe^{3+}$$

The oxidation in appropriate ranges of radiation dose is proportional to the amount of radiation. The number of ferrous ions oxidized per 100 eV of energy absorbed, designated as the G value, is 15.6.

The ferrous sulphate-cupric sulphate dose meter is a modification of the Fricke ferrous sulphate dose meter. The G value of 15.6 limits the Fricke dose meter to a dose range of 20 to 400 Gy (2 to 40 krad). Adding cupric sulphate to the Fricke dose meter reduces G (Fe^{3+}) to 0.66. The ferrous sulphate-cupric sulphate dose meter has a range of 100 Gy to 8 kGy (10 to 800 krad). This range is obtained when the dose-meter solution is an air-saturated solution containing 0.001M $FeSO_4$, 0.010M $CuSO_4$ and 0.01N H_2SO_4. The range can be extended to above 30 kGy (3 Mrad) by modifying the solution so that the ferrous ion concentration is 0.01M. This is shown by the data in Fig.8.

The reactions of the ferrous sulphate-cupric sulphate dose meter are as follows:

$$Fe^{2+} + OH = Fe^{3+} + OH^-$$

$$H + O_2 = HO_2$$

$$Cu^{2+} + H (e^- aq) = Cu^+ + H^+$$

$$Cu^{2+} + HO_2 = Cu^+ + H^+ + O_2$$

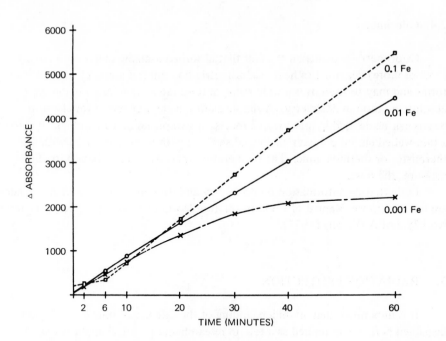

FIG.8. Effect of initial ferrous ion concentration on linear portion of absorbance-irradiation time (dose) relationship. Dose rate approximately 600 Gy/min (60 krad/min). Reprinted from Radiation Preservation of Foods, Advances in Chemistry, Ser.65 (1967) 82.

$$Fe^{3+} + Cu^+ = Fe^{2+} + Cu^{2+}$$

$$2 Fe^{2+} + H_2O_2 = 2 Fe^{3+} + 2 OH^-$$

There is no dose-rate effect, since all but the last reaction are fast. This reaction has a half period of 14 s. Since oxygen is not consumed, the $G (Fe^{3+})$ value is not affected.

The need for simple ready-made dose meters, suitable for routine use, has led to the development of a number of plastic film systems. Examples of such films for routine dosimetry are the red Perspex and clear polymethylmethacrylate. Change in optical transmission value is measured spectrophotometrically. Such systems have proved extremely useful for dose-distribution measurements and for process monitoring. Careful calibration must be carried out for each production batch of such plastic films, often even for each sheet of material. This calibration can be done by checking against the Fricke dose meter, which is always used as the standard.

2.4. Calorimetry

Calorimetry depends on the fact that absorbed ionizing radiation is largely (95% or more) degraded to heat. Calorimetric dose meters are available in several forms and may be used over a wide range of dose rates; they provide the only absolute method in dosimetry. A simple calorimetric dose meter for electron beams can be devised by placing in a recess, in two pieces of 10-cm-thick styrofoam, a thin-walled closed polystyrene Petri dish filled with water and fitted with a thermistor or thermocouple. The temperature increase obtained upon irradiation measures the dose.

For extensive information on dosimetry and dose meter measurements readers are referred to the Manual of Food Irradiation Dosimetry, Technical Reports Series No.178, IAEA, Vienna (1977).

3. RADIATION PROTECTION

It is imperative that an understanding of the safe use of ionizing radiation be gained before it is applied as a tool in research and practical applications to food irradiation. Ionizing radiation is hazardous to all biological systems but, with proper attention to health protection measures, the hazard to personnel can be reduced to a tolerably low level. The health physics involved in the safe use of ionizing radiation is discussed in some detail under three headings:

Protection of personnel;
Control and monitoring of external radiation hazard;
Radiation protection of large irradiation facilities.

Before considering these categories, however, an insight must be gained into the basics of radiation protection and units.

3.1. Basic considerations and units

A health hazard is involved when human tissues are subjected to ionizing radiation. The effects of radiation on the human body are the result of damage to individual cells. At the cellular level the best-known effects are mutations, chromosomal aberrations and cell-reproductive death. The severity of the effect depends on many factors, such as absorbed dose, dose rate, type of radiation, cell type, cell cycle, metabolism (repair capacity), conditions (O_2, N_2, sensitizers, etc.).

It is clear, however, that radiation dissipating 1 Gy (100 rad) with a high specific ionization, i.e. a high linear energy transfer (LET), will have a greater biological effect on an organism than a different quality of radiation dissipating 1 Gy with a low specific ionization.

To allow comparison of effects of different radiation qualities, a special unit is introduced: this is *dose equivalent*, the units of which are the *rem* and the *sievert* (the sievert is a new SI unit).

Dose equivalent (rem) = absorbed dose (rad) $\times$ Q
Dose equivalent (sievert) = absorbed dose (Gy) $\times$ Q
1 Sv (1 sievert) = 100 rem

TABLE III. VALUES OF QUALITY FACTOR Q, USED IN DEFINING DOSE EQUIVALENT

Radiation	Q
X-, γ-, β-rays, electrons	1
Thermal neutrons	2.3
Fast neutrons	10

Q = quality factor, which reflects the ability of a particular kind of ionizing radiation to cause damage. When expressed in dose-equivalent units, the damage of different radiations is similar. Some data concerning quality factor Q are shown in Table III.

The effects of radiation on the human body can be divided into three classes:

(a) Acute effects

Below a threshold value of 50 rem no effects are observed. Above this value, the severity of the acute effects is dose-dependent. Characteristics of acute effects are: aberrations in blood composition (e.g. lymphocytes); general illness; loss of hair; vomiting; lethality occurring at doses of 500 to 1000 rem. These symptoms are observed within three months following exposure.

(b) Late somatic effects

The latent period of late somatic effects varies between 5 and 20 years. There is no threshold value. The chance of late somatic effects increases with dose equivalent accumulated over many years. The symptoms of late somatic effects are leukaemia, cancers and tumours. The severity of the disease is dose-independent.

(c) Genetic effects

Genetic effects manifest themselves in the next generations. There is no threshold value, and the chance of effect increases with dose equivalent accumulated before conception.

3.2. Protection of personnel

Throughout his history man has been exposed to irradiation from the environment in which he lives. The three main natural sources are:

Cosmic radiation: Dependent on altitude. At sea level, 50 mrem·a^{-1} (0.5 mSv·a^{-1}).

Terrestrial radiation: Radioactive elements in rocks and soil (uranium, thorium, radium). Dose rate: 30 to 60 mrem·a^{-1} (0.3 to 0.6 mSv·a^{-1}).

Radioactivity in the body: Inhaled and ingested radionuclides (^{14}C, ^{40}K, radon, thoron). Dose rate: 20 mrem·a^{-1} (0.2 mSv·a^{-1}).

Total contribution from natural radiation sources is approximately 125 mrem·a^{-1} (1.25 mSv·a^{-1}).

In addition to the natural sources, man is exposed to 'man-made' sources of radiation, which include:

Diagnostic radiology (X-rays);
Therapeutic radiology (large exposures, but small number of people involved);
Fall-out products (^{90}Sr, ^{137}Cs);
Use of radionuclides (industry, research, agriculture);
Nuclear power;
Irradiation facilities.

Table IV estimates the proportional distribution between different radiations from man-made sources.

The International Commission on Radiological Protection (ICRP), Sutton (Surrey, UK), regularly produces maximum permissible doses for radiation workers (occupational limits) and for man (members of public). The recommendations of the ICRP since 1969 are summarized in Table V.

In the case of whole-body irradiation, the gonads and red bone marrow are considered to be critical. The limit for whole-body irradiation to radiation workers must not exceed 5 rem per year (= 0.05 Sv·a^{-1}). The annual maximum of 5 rem applies to the dose from both natural and man-made (external exposure) radiation

26

TABLE IV. AVERAGE DOSE TO POPULATION

Sources	Dose per year	
	mrem	μSv
Diagnostic radiology	70	700
Therapy	5	50
Fall-out	4	40
Radiological workers	1	10
Other sources	3	30
Nuclear power (1970)	0.003	0.03
Nuclear power (2000)	1	10

TABLE V. MAXIMUM PERMISSIBLE DOSES

Organ	Occupational limits	Member of public
Gonads, red bone marrow	$5 \text{ rem} \cdot a^{-1} = 0.05 \text{ Sv} \cdot a^{-1}$	$0.5 \text{ rem} \cdot a^{-1} = 5 \text{ mSv} \cdot a^{-1}$
Skin, bone, thyroid	$30 \text{ rem} \cdot a^{-1} = 0.3 \text{ Sv} \cdot a^{-1}$	$3.0 \text{ rem} \cdot a^{-1} = 30 \text{ mSv} \cdot a^{-1}$
Hands, forearms, feet, ankles	$75 \text{ rem} \cdot a^{-1} = 0.75 \text{ Sv} \cdot a^{-1}$	$7.5 \text{ rem} \cdot a^{-1} = 75 \text{ mSv} \cdot a^{-1}$
Other single organs	$15 \text{ rem} \cdot a^{-1} = 0.15 \text{ Sv} \cdot a^{-1}$	$1.5 \text{ rem} \cdot a^{-1} = 15 \text{ mSv} \cdot a^{-1}$

sources. All unnecessary exposures should be avoided. If exposures are inevitable, the doses should always be kept as low as reasonably possible, but in any case below the limits given in Table V.

3.3. Control of external radiation hazard

The external radiation hazard is controlled by three factors: time, distance and shielding.

27

(a) Time

Reduction of exposure time is important in minimizing dose. Manipulations with sources should be performed rapidly but carefully. From

Dose = dose rate × time

it follows that the maximum permissible dose-rate level for occupational work on the basis of 50 weeks per year and 40 hours per week must not exceed $2.5 \text{ mrem} \cdot \text{h}^{-1}$ (= $25 \ \mu\text{Sv} \cdot \text{h}^{-1}$, or 1000 mrem per week) (= 1 mSv per week).

(b) Distance

Distance is a very important factor in minimizing dose. For point sources of activity, γ-ray intensity is inversely proportional to the square of the distance. Thus, once the exposure dose is known at any one distance, it may be calculated at any other distance by the inverse-square law. Consider a point source from which the γ-ray exposure dose was 1 mrem per hour at 10 cm. Any manipulations of the source by means of long forceps or tweezers would produce a negligible finger or whole-body dose. However, if the source were handled without tweezers (for instance with rubber gloves as the only protection), the radiation exposure dose at 1 mm distance would be 10 rem per hour to the skin of the fingertips.

$$\text{Dose rate (D) for } \gamma\text{-sources} = 0.5 \times \frac{CE}{r^2} \text{ rem} \cdot \text{h}^{-1}$$

where

C = activity of the source in curies;
E = total γ-energy per disintegration in MeV;
r = distance to the source in metres.

In SI units:

$$D = 1.4 \times 10^{-13} \times \frac{CE}{r^2} \text{ Sv} \cdot \text{h}^{-1} \text{ with C in Bq}$$

(c) Shielding

Gamma radiation is much more penetrating than α- and β-radiation because it has a low LET. The absorption of electromagnetic radiation is exponential. A beam of photons can therefore never be completely stopped by matter although the transmitted radiation may be reduced to an insignificant value. The attenuation of monoenergetic photons by absorption and scattering is described in detail in Section 1.4.

28

TABLE VI. THICKNESS IN cm TO DECREASE THE RADIATION DOSE RATE
BY VARIOUS FACTORS (^{60}Co γ-RADIATION)

Attenuation factor Shield material	10	10^3	10^6
Lead (density 10.8)	5	13.3	25.4
Iron (density 7.8)	9.2	22.8	41.6
Heavy concrete (density 3.4)	20	51	92
Ordinary concrete (density 2.3)	32	75	135
Water (density 1.0)	70	145	850

In the case of γ-rays, a high-Z material such as lead, usually applied in 'self-contained' irradiators for food irradiation research, provides the best shielding. In this case, the source material (^{60}Co) is contained in a cask surrounded by a lead shield; the food sample material to be irradiated is inserted in a 'drawer' and brought into the presence of the radiation field. This is done by manipulation in such a way that *at no time* does the irradiation escape from the irradiator into the space occupied by the worker.

In certain cases, a small room is made suitable for irradiation by shielding it properly, and the irradiation source material is stored in a safe position, i.e. under water or beneath a lead plug. The source is manipulated remotely, so that samples left in the room will be irradiated when the source is brought from the 'safe' storage position into the irradiating position. By means of safety-locking devices, the room can be secured so that it can be entered by the worker *only* when the source is in the safe position.

In pilot, semi-industrial and commercial irradiators, concrete is the usual shielding material of the irradiation chamber. In each of these cases, the adequacy of the shielding is very important, since the workers must be adequately protected against irradiation at all times. Any radiation source used in food irradiation work must naturally comply with all the safety regulations covering industrial installations and must be safe from the point of view of radiation. Table VI gives the thickness of various materials required to decrease the dose rate for a ^{60}Co source by factors of 10, 10^3 and 10^6. This table can be used as a guide in estimating the amount of shielding necessary when planning the construction of a radiation facility. Work on and in the vicinity of radionuclide sources must always be performed with sufficient shielding for personnel.

3.4. Monitoring of external radiation hazard

Areas surrounding irradiation sources must be checked routinely (radiation survey monitoring) to test the adequacy of shielding and to demonstrate that radiation levels are satisfactory.

A survey monitor should (a) be capable of monitoring radiation, (b) be portable, (c) be easy to use, (d) indicate dose rate in $rem \cdot h^{-1}$ ($mSv \cdot h^{-1}$) and (e) be equally sensitive to all radiation energies.

The following devices are commonly used:

Ionization chamber (good energy response; not sensitive).
Gas proportional counters.
Geiger-Müller tube (poor energy response; very sensitive).
Scintillation counter (poor energy response; very sensitive).

It is imperative to measure the accumulated dose to radiological workers and source operators (personnel monitoring). Any person dealing with or working in the vicinity of radioactive isotopes should wear personal dose meters. These can be worn on the body, or attached to the hands or wrists if necessary. They provide an integrated dose reading, i.e. a dose value summed over the total working period. Personal dose meters commonly used are:

Film badge: minimum dose − 10 mrem; maximum dose − 1000 rem
(0.1 mSv to 10 Sv).
Pen dose meters: pocket ion chamber with a direct visual indication
of the accumulated dose.
Pocket alarm dose meters.
Thermoluminescent dose meters: solid-state dose meter which will
replace film badge in future.

3.5. Radiation protection of large irradiation facilities

In food irradiation, the size of the source and the radiation energy involved make it absolutely necessary to depend on shielding as the safety factor against exposure to ionizing radiation. The main factors in radiation protection at irradiation facilities are:

(a) Prevention of overexposure under normal operation

The main measures usually taken are:

Interlocked entrance; no admission when source is operating.

Audible alarm when source is made operational.

Scram buttons in case of emergency.

After operation: entry into irradiation room only with monitor.

(b) Low radiation levels at places admissible to members of the public

Radiation level outside the irradiation chamber must not exceed 0.25 mrem·h^{-1} (2.5 μSv·h^{-1}). Critical points are: entrance door to the irradiation chamber; food-products discharge gates; holes in the concrete walls (shielding) for electric wires, etc.; and the roof.

Areas with a radiation level higher than 0.25 mrem·h^{-1} must be designated as restricted areas. A warning sign: 'Caution — high-radiation area' should be used.

(c) Contamination control

Comprises tests for contaminated water and smear tests of radiation rods. Permissible leakage of encapsulated sources: 5 to 50 nCi ($\sim$200 to 2000 Bq).

(d) Reduction of ozone risk

Large sources produce considerable amounts of ozone. The maximum permissible level for persons is 0.1 cm^3·m^{-3}. The irradiation room should be ventilated. Delay in entering irradiation room is sometimes necessary.

4. RADIATION CHEMISTRY

4.1. Fundamentals

The use of ionizing radiation in food irradiation is limited to electromagnetic radiation of energy not greater than 5 MeV and to electrons of energy not greater than 10 MeV.

When γ-rays or X-rays are absorbed, the most important mechanism is Compton scattering (see Section 1.4.3). Part of the photon energy is used to eject an electron from an atom of the absorber while the remainder is retained. The diminished-energy photon may repeat the process with another atom. The ejected electron, carrying a large portion of the original energy of the photon, also loses energy through ionization and excitation of the molecules of the absorber. Ultimately both the photon and electron energies are reduced by repetition of these processes to essentially zero values.

Electron irradiation operates similarly, producing ionizations and excitations along a path or 'track' of progress through the absorber.

Two basic processes occur when ionizing radiation acts upon matter. The primary process causes the formation of ions, excited molecules or molecular fragments.

Ionization: $M \xrightarrow{\hspace{0.8cm}} M \cdot^+ + e^-$

Dissociative ionization: $M \xrightarrow{\hspace{0.8cm}} A^+ + B \cdot + e^-$

Excitation: $M \xrightarrow{\hspace{0.8cm}} M^*$

Dissociation: $M \xrightarrow{\hspace{0.8cm}} A \cdot + B \cdot$

The secondary process involves interaction of the products of the primary process and can lead to the formation of compounds different from those initially present. While the primary process is independent of the temperature, the secondary process is dependent on temperature and other variables (e.g. for gases, pressure). For the interaction to take place, the primary products must come together. This requires sufficient time and the absence of constraints. Reactions involving the secondary products may also occur. Types of reactions of the secondary process are given below.

Electron capture:
$$M + e^- \rightarrow M \cdot^-$$
$$\rightarrow A \cdot + B^-$$
$$\rightarrow M \cdot^{-*}$$

Ion-molecule reactions:
$$A^+ + M \rightarrow C^+ + D \cdot$$
$$\rightarrow C^{+*} + D \cdot$$
$$\downarrow$$
$$E^+ + F$$
$$\downarrow M$$
$$G^+ + H$$

Excited molecules:
$$M^+ + e^- \rightarrow M^*$$
$$A^+ + B^- \rightarrow AB^*$$
$$M^* + m \rightarrow M + m*$$
$$M^* \rightarrow M + h\nu$$
$$M^* \rightarrow A \cdot + B \cdot$$
$$M^* \rightarrow C + D$$

Radical reactions:
$$A \cdot + m \rightarrow Am \cdot$$
$$A \cdot + nm \rightarrow An + m \cdot$$
$$A \cdot + B^- \rightarrow A^- + B \cdot$$
$$A \cdot + B \cdot \rightarrow AB$$
$$2A \cdot \rightarrow C + D$$

It has been noted in Section 1.4 that the interaction of ionizing radiation with matter is complex. Nonetheless, the resultant effect at the molecular level for both electromagnetic and corpuscular radiation is as given above if the limits of energy are below 10 MeV. Above this value, nuclear transformations, leading in some cases to induced radioactivity, also become probable.

Although the incident ionizing radiation may be monoenergetic, the actual radiation within the absorbing medium encompasses a wide range of energies. As the incident radiation proceeds into the medium, it is dissipated, primarily by interaction with electrons. A sufficiently energetic interaction can lead to the ejection of an electron from an atom or molecule and cause ion formation. A less energetic collision may cause only the transfer of energy to the atom or molecule to cause excitation. By repetition of these processes a degradation of the energy of the incident radiation occurs. In this way, along the path of the radiation exists a range of energies. Some of these different energy levels are associated directly with the primary radiation, and some are those related to secondary radiation produced by interaction of the primary radiation with the absorber.

It is to be noted that only absorbed radiation can accomplish change. The G-value is the number of molecules reacting or produced per 100 eV of absorbed energy. The number of reacting molecules per unit of radiation can vary greatly. Many activated molecules lose their energy before a chance for reaction occurs, and in this way cause some inefficiency in the process. In some cases, an exothermic reaction occurs, causing excitation of additional molecules and leading to additional reactions. Chain reaction can also be initiated; in this way, a small amount of radiation can cause substantial change.

4.2. Reactions relating to non-foods

Irradiation of gaseous O_2 can lead to ozone formation. Mixtures of nitrogen and oxygen form nitrogen oxides, which in the presence of water form nitric acid. Ozone is also formed. These reactions may be relevant to the irradiation of certain foods in the presence of air or oxygen, since ozone and nitrogen oxides can cause changes in foods through chemical action.

One of the most interesting and important substances which undergo change when irradiated is water. The final products of the irradiation of water are only hydrogen and hydrogen peroxide, but the mechanism whereby these are formed is known to be complex.

The effects of radiation on pure water and dilute solutions can be expressed in a simplified manner by the equation:

$$H_2O \rightarrow 2.7\ OH\cdot + 2.7\ e_{aq}^- + 0.55\ H\cdot + 0.45\ H_2 + 0.71\ H_2O_2 + 2.7\ H_3O^+$$

in which the numbers are G-values.

The hydroxyl radical $OH\cdot$ is derived chiefly from positive ions formed by ionization of water:

$$H_2O\cdot^+ + H_2O \rightarrow H_3O^+ + OH\cdot$$

The hydrated electron e_{aq}^- is formed from electrons freed during ionization.

Hydrogen atoms $H\cdot$ are obtained partly by dissociation of excited water molecules and partly from reactions of hydrated electrons with hydrogen ions:

$$e_{aq}^- + H_3O^+ \rightarrow H_2O + H\cdot$$

Hydrogen atoms form molecular hydrogen:

$$2\,H\cdot \rightarrow H_2$$

Similarly, hydroxyl radicals form hydrogen peroxide:

$$2\,OH\cdot \rightarrow H_2O_2$$

Pure water, however, does not actually decompose when irradiated. The end molecular products given above, H_2O_2 and H_2, are themselves consumed by the radicals formed:

$$e_{aq}^- + H_2O_2 \rightarrow OH\cdot + OH^-$$

$$OH\cdot + H_2 \rightarrow H_2O + H\cdot$$

The presence of a solute can alter the situation. Dissolved O_2, for example, by reacting with e_{aq}^-, causes the formation of H_2O_2, as follows:

$$e_{aq}^- + O_2 \rightarrow O\cdot_2^-$$

$$H^+ + O\cdot_2^- \rightarrow HO\cdot_2$$

$$O\cdot_2^- + HO\cdot_2 \xrightarrow{+H^+} H_2O_2 + O_2$$

$$2\,HO\cdot_2 \rightarrow H_2O_2 + O_2$$

Other solutes can also alter the end results of irradiation of water.

As might be expected, the conditions under which the irradiation is conducted affect the results. Temperature, pH value and purity of water can be influential.

While the end products, H_2 and H_2O_2, are of interest, the intermediate ones can also be important. Two points are especially noteworthy:

(1) While transient, these intermediates do exist for a finite period; and

(2) In at least some of the possible physical states of water, the active molecular fragments can move and in this way contact other molecules with which they can react.

For aqueous solutions or for materials containing water, such as foods, these highly reactive products of the radiolysis of water can produce an indirect effect of the radiation with the substances dissolved in the water.

Because the products of the radiolysis of water include substances which are either oxidants or reducing agents, it is clear that both oxidations and reductions can occur upon the irradiation of water. The Fricke dose meter discussed in Section 2.3 is an example of an oxidation induced by radiation.

The radiolysis of organic compounds can also be complex. A number of primary processes occur, leading to excitation, ionization and dissociation. The possibilities of secondary effects from the interaction of these products can be manifold, depending on the chemical nature of the original material and the conditions of irradiation. There may be a pattern of somewhat random splitting of the chemical bonds present, as occurs, for example, with saturated hydrocarbons. On the other hand, certain bonds are more readily broken than others. It is therefore impossible to predict the exact outcome of the irradiation of a particular compound or a mixture of compounds. One can find fragments of lower molecular weight than the original substance, polymers, products of the interaction of radiation-produced intermediates with each other or with parent substances, altered parent substances (e.g. dehydrogenation, decarboxylation, de-amination, etc.), and new compounds formed by interaction of two or more starting substances (e.g. the formation of benzene hexachloride from Cl_2 and benzene). Exothermic reactions or chain reactions in some cases produce high efficiencies in the use of radiation to produce changes.

While the complexity of the radiation-induced reactions makes prediction of the course of a particular reaction difficult, control of conditions does lead to consistent results, and useful applications of radiation exist. For example, cross-linking of polyethylene to raise the melting point of this material is carried out commercially.

4.3. Reactions relating to foods

The radiation chemistry of food components is both interesting and important since it helps in the understanding of what happens to actual foods when they are irradiated. In this section, attention is directed to the macrocomponents (proteins, carbohydrates and lipids), and only to those microcomponents (the vitamins) which are potentially susceptible to change through irradiation.

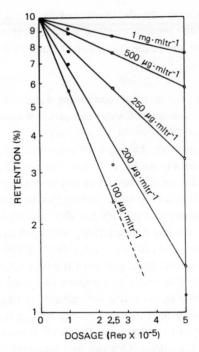

FIG.9. Decomposition of DL-phenylalanine by high-voltage cathode rays. (To convert Rep to Gy, multiply by 0.0093.)

4.3.1. Amino acids and proteins

The study of amino acids can provide information that helps in the understanding of changes in the very much more complex protein molecules. The principal radiolytic changes in aqueous solutions of the simple aliphatic amino acids are de-amination and decarboxylation. A number of products result, including NH_3, CO_2, H_2, amines, aliphatic acids and aldehydes. Sulphur-containing amino acids are likely to be more sensitive to radiation. Usually the sulphur moiety is oxidized, H_2S, elemental S or volatile sulphur compounds being formed.

Cysteine, containing the -SH group, forms the disulphide cystine. Amino acids with a ring structure can undergo rupture of the ring. De-amination of aromatic amino acids occurs but not to the same degree as with aliphatic amino acids. Hydroxylation of the aromatic ring is the principal reaction of phenylalanine and of tyrosine. The indirect effect of radiation increases with decreasing concentration, as shown in Fig.9 for the decomposition of DL-phenylalanine in water by electrons.

A protein molecule responds to radiation in a dual manner: as a protein entity and as individual amino acids and other constituents. In the dose ranges used

36

in food irradiation the effect on proteins is usually small and very much the same, regardless of the origin of the protein. To some extent, the particular effect of radiation is related to the protein's structure, its composition, whether native or denatured, whether dry or in solution, whether liquid or frozen, and to the presence or absence of other substances. The generalized phenomenon of denaturation by irradiation manifests itself in changes commonly associated with the kind of protein alteration, i.e. changes in viscosity of solutions, in solubility, in electrophoretic behaviour, in changes in absorption spectra, in reactions with enzymes, in exposure of -SH groups and in immunological changes. Splitting of protein molecules into smaller units can occur. Aggregation has also been noted, and both fragmentation and aggregation have been observed to occur simultaneously.

Most proteins are large molecules. Because of this, they can provide more than one locus for radiation-induced changes. Absorbed energy can be translocated to a more sensitive 'site', where an initially broken bond can lead to specific chemical changes such as were indicated for isolated amino acids. In proteins containing sulphur, the sensitive location is at the sulphur linkage. For a particular protein a characteristic set of radiation-sensitive sites can exist, and this results in a preferred and consistent response, in contrast to a random and varied effect.

In terms of the protein as a whole, the consequences of the changes just described can be varied; for example, denaturation, colour changes (in chromo-proteins) or impairment or alteration of functional characteristics in biological systems or in food uses. In biological systems such changes can be lethal to an organism.

All these effects result from a combination of the primary (direct) and secondary (indirect) processes referred to in Section 4.1. The relative importance of each kind of effect will vary, depending on many factors, such as concentration, availability of oxygen, temperature, nature of the protein, and the presence of other substances.

Irradiation of dry proteins, however, involves almost entirely the primary or direct effect. As with amino acids, free radicals formed from a variety of moieties of the protein molecule have been observed in dry proteins.

The value of food proteins as nutrients largely relates to their component amino acids. As noted above, radiation can alter amino acids. The doses employed in food irradiation, however, do not result in significant changes in the amino acid composition of protein foods; consequently irradiation causes no measurable nutritional loss of protein value.

4.3.2. Enzymes

Enzymes are important constituents of living tissue, and since many foods are either living organisms (e.g. fresh fruits) or are closely derived from such (e.g. fresh meat), enzymes are frequently constituents of the foods. There are a great many

TABLE VII. ELECTRON INACTIVATION OF PEPSIN IN AQUEOUS SOLUTION

mg pepsin/mltr	Gy to produce 63% inactivation
0.5	1360
1.0	2600
2.0	5230

Data from J. Food Technol. **6** (1952) 239.

enzymes and all are proteins. Since they are proteins, it is to be expected that the action of radiation on enzymes will be no different from that on other proteins. Enzymes, however, exhibit certain specific functional characteristics which are frequently of interest in connection with a food. Changes in these characteristics, or sometimes a failure to respond to radiation, can therefore be a matter of special concern. The enzyme activity is usually a sensitive and convenient index of change, making it possible to detect the action of radiation on these proteins more easily than may be the case for some other proteins.

As is to be expected, the circumstances under which the enzyme exists have a great influence on the changes induced by radiation. Inactivation appears to come about through denaturation, but other more specific changes can also occur. Dilute pure aqueous solutions of enzymes are sensitive to radiation. Increasing the concentration requires more radiation to produce the same inactivation, as may be seen from the data of Table VII.

The inclusion of other substances in the enzyme solution also decreases the sensitivity of the enzyme to radiation. Figure 10 gives data showing the protective action of sodium D-isoascorbate on pepsin in acetate buffer solution. The radiation sensitivity of enzymes in aqueous solution increases with temperature, as shown for pepsin in Fig. 11.

Other factors affect the sensitivity of enzymes to radiation. Enzymes that are dependent on the presence of an -SH group for activity are especially sensitive. The pH value of the solution, or its oxygen content, is an important factor for some enzymes. Inactivation of dry enzymes requires greater amounts of radiation than aqueous solutions.

The sensitivity of enzymes to radiation is therefore not simply stated. Details of the environment of the enzyme must be known and generally it can be expected that the more complex this environment, the less sensitive to radiation will the enzyme be. In the usual complex food systems, enzymes are well protected and the radiation requirements for inactivation are quite large.

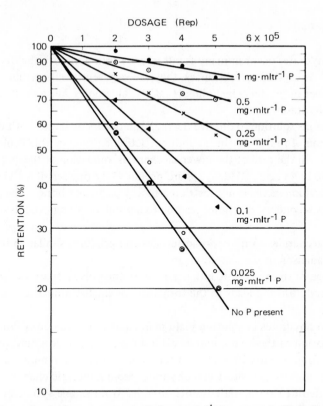

FIG.10. *Effects of cathode rays on pepsin (1.0 mg·mltr⁻¹) in acetate buffer (pH 4.3) in the presence of different amounts of sodium D-isoascorbate. P is protector. (To convert Rep to Gy, multiply by 0.0093.) From J. Food Technol.* **6** *(1952) 239.*

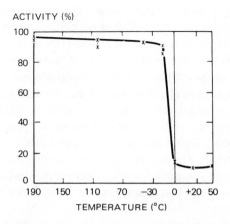

FIG.11. *Effect of temperature on the inactivation of 10 mg·mltr⁻¹ solution of pepsin by 23.25 kGy (2.325 Mrad). From Ann. N.Y. Acad. Sci.* **59** *(1955) 595.*

4.3.3. Carbohydrates

Carbohydrates occur in many foods and are found in both the dry and wet conditions. As crystalline substances, they are very sensitive to radiation, yielding a large number of products including H_2, CO_2, aldehydes, ketones, acids and other carbohydrates.

In aqueous solution, oxidative degradation occurs owing to both direct and indirect action of radiation. In the indirect action a principal role is played by the OH· radical. In the case of the lower saccharides, oxidation at the ends of the molecule produces acids. Aldehydes are produced by ring scission. The action of radiation on aldohexoses is not restricted to any particular bond. The presence of O_2 causes secondary reactions, leading to a number of compounds including glyoxal.

Oligosaccharides form monosaccharides and products similar to those obtained with the irradiation of the simple sugars.

Cleavage of the glycosidic link occurs with the polysaccharides such as starch and cellulose and leads to the formation of smaller units such as glucose, maltose, etc.

All carbohydrates in solution yield malonaldehyde and deoxy compounds. In the formation of these compounds pH is an important factor, and the normal pH of most foods greatly limits this production. Proteins and amino acids as well as the other substances protect carbohydrates from radiolytic change. Because of this it is difficult to extrapolate results obtained from simple systems to what occurs in complex systems such as those that exist in foods.

4.3.4. Lipids

At doses below 50 kGy (5 Mrad) the changes in common indices of fat quality are slight. Undesirable flavour changes occur, however, at doses as low as 20 kGy (2 Mrad). At higher doses up to 1 MGy (100 Mrad) there can be significant changes in both chemical and physical characteristics.

Two types of change induced in fats by irradiation can be identified: autoxidative and non-oxidative. The autoxidative process induced by irradiation is much the same as that which occurs without irradiation. Irradiation, however, accelerates the process. It produces free radicals whose types and decay rates are affected by temperature. After irradiation these free radicals can react with O_2 over an extended period. The free radicals cause the formation of hydroperoxides, which yield a variety of compounds including alcohols, aldehydes, aldehyde esters, hydrocarbons, hydroxy and keto acids, ketones, lactones, oxoacids and dimeric compounds.

The non-oxidative changes occur when O_2 is excluded during and after irradiation. Radiolytic products include H_2, CO_2, CO, hydrocarbons and aldehydes.

With unsaturated fats, some hydrogenation occurs and significant amounts of dimers are formed. The general mechanism for the non-oxidative radiolysis of a triglyceride involves cleavage at five preferred locations in the molecule and randomly at the remaining carbon-to-carbon bonds of the fatty acids, as shown below. (The preferred locations are at a, b, c, d and e; n is the number of carbon atoms in the fatty acid.)

$$\begin{array}{c} \overset{a}{\underset{}{\text{C}}} \!\!-\!\! \text{O} \!\!-\!\! \overset{b}{\underset{}{\text{C}}} \!\!-\!\! \overset{\overset{\text{O}}{\|}}{\text{C}} \!\!-\!\! \overset{c}{\underset{}{\text{C}}} \!\!-\!\! \overset{d}{\underset{}{\text{C}}} \!\!-\!\! \text{C} \!\!-\!\! \text{C} \!\!-\!\! \text{R} \\ e \\ \text{C} \!\!-\!\! \text{O} \!\!-\!\! \text{O} \!\!-\!\! \text{C}_n \\ | \\ \text{C} \!\!-\!\! \text{O} \!\!-\!\! \text{O} \!\!-\!\! \text{C}_n \end{array}$$

The free radicals formed by scission in this manner mainly add hydrogen obtained by extraction from other molecules or, to a lesser extent, they either lose hydrogen or combine with other free radicals. In this way a number of stable radiolytic substances are formed. The particular compounds formed relate to the initial composition of the lipid. The amounts for a given lipid are in proportion to dose. As measured by volatile substances detected, the radiolytic products are produced in quite small quantities. For example, the volatiles found in beef fat irradiated with 60 kGy (6 Mrad) total less than 0.03%. Irradiation in the solid or liquid state affects the relative amounts.

While there are some differences, the effects of radiation on lipids are similar to those of heat. The compounds responsible for the 'irradiated flavour' of fat have not been identified.

4.3.5. Vitamins

Vitamins are important micronutrients in foods. Many food preservation processes cause some vitamin loss, and the effect of irradiation on these substances is a reasonable matter of interest. The structures of most vitamins are known. Vitamins can usually be obtained in pure forms, often synthetically produced. As would be expected, the effect of radiation on a vitamin is greatly dependent on the environment in which the vitamin exists. Simple systems such as a pure water solution, especially if dilute, show a large effect of radiation. More complex environments such as exist in foods lead to a reduced sensitivity to radiation.

Vitamins can be classified as to whether they are water- or fat-soluble. The widely varying chemical structures of vitamins result in differences in response to radiation. Of the water-soluble vitamins, Vitamin B_1 (thiamine) is the most sensitive. The chemical structure of B_1 is such as to make it susceptible to attack by e_{aq}^-. Vitamin C (ascorbic acid), another water-soluble vitamin, is also radiation-sensitive and forms dehydroascorbic acid and other products. The effect of radiation depends on the concentration of this vitamin in water, as shown in Fig.12(a).

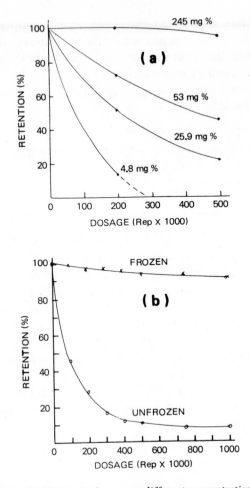

FIG.12. (a) The effect of 3-MeV cathode rays on different concentrations of L-ascorbic acid.
(b) The effect of 3-MeV cathode rays on frozen and unfrozen solutions of L-ascorbic acid (8.0 mg%) in 0.25% oxalic acid.
(To convert Rep to Gy, multiply by 0.0093.)

Figure 12 (b) indicates the effect of freezing on the retention of ascorbic acid in aqueous solution. Both of these effects demonstrate the indirect action of radiation. Other water-soluble vitamins sensitive to radiation are riboflavin, Vitamin B_{12} and biotin. Niacin, pantothenic acid and folic acid appear to be quite resistant.

Of the fat-soluble vitamins, Vitamin E is the most radiation-sensitive. Vitamin A and carotenoids, and Vitamin D, likewise undergo radiolytic change. Vitamin D is resistant in the dose range below 50 kGy (5 Mrad).

The radiation sensitivities indicated above should not be regarded as adequately describing the retention of a particular vitamin in an irradiated food. The dose employed, the conditions of irradiation, and the environment in which the vitamin occurs can greatly alter its stability.

5. EFFECTS OF RADIATION ON LIVING ORGANISMS

5.1. Fundamental concepts

It is presumed that the biological effects of radiation are due to chemical changes within the organism. As with other materials, the effects of radiation on living organisms can be divided into direct and indirect effects. The presence of substantial quantities of water is normal in living tissue. Consequently the indirect effect of radiation normally occurs as an important part of the total action of radiation. Drying or freezing of the tissue can reduce or remove this mechanism.

The effect of radiation on a living organism requires a certain time for its manifestation. The sequence of events following irradiation can occur in different ways depending on the dose. Radiation damage is mainly associated with the impairment of metabolic reactions. An important aspect of the reaction of living organisms to radiation is the capability of the organism to recover from radiation injury. This capability is related to many factors, perhaps the most important of which is total dose. A sufficiently high dose can prevent recovery.

The effects of radiation cannot be simply described for all organisms, since these effects are related to the nature of the organism and especially to its complexity. The correlation of radiation sensitivity is roughly inversely proportional to size. The viruses, the most minute living entities, are most radiation-resistant, some surviving as many as 100 kGy (10 Mrad). Man, approximately at the other end of the size range and complexity, suffers death with only 5 Gy (500 rad). Effects on other types of organisms fall within these limits, as shown in Table VIII.

Since all living organisms (except viruses) contain one or more cells as the basic unit, the effects of radiation on cells have been studied extensively. Effects which have been detected include changes in shape and structure, alteration of metabolic reactions, alteration of reproduction, including mutations, altered nutrient requirements and death. Cells in an active metabolic state are more sensitive to radiation than those in a dormant or resting condition.

The consequence of radiation damage to a cell will vary with the nature of the organism. Single-cell organisms are more vulnerable than multicell organisms to the consequences of radiation damage, since the cell is the whole organism. Damage to some of the cells of a multicell organism does not necessarily have serious consequences for the total organism. On the other hand, the complexity

TABLE VIII. APPROXIMATE DOSE TO KILL VARIOUS
ORGANISMS

Organism	Dose (Gy)[a]
Higher animals, including mammals	5–10
Insects	10–1000
Non-sporulating bacteria	500–10 000
Sporulating bacteria	10 000–50 000
Viruses	10 000–200 000

[a] To convert to rad, multiply by 100.

of a multicell organism can make certain kinds of damage critical for its functioning. This explains in part the inverse relation between lethal dose and complexity of the organism.

For doses less than lethal the effects of radiation will vary depending on the timing of the irradiation with respect to the stage of development of the organism. Application during the growth stage, for example, can affect the maturation of the organism and could include alteration of the normal structure, metabolism, reproduction, etc.

5.2. Actions on organisms relating to foods

Living organisms are associated with foods in various ways. Some foods, such as fruits and vegetables, are themselves living organisms. The effects of radiation on these can be significant and useful in the various steps of handling foods between harvest and consumption.

The presence of a living organism in a food can affect the acceptability of that food. The significance of the presence of the organism can vary from a true health hazard to objection on the basis of aesthetic or spoilage changes of a sensory character. The significance of a particular organism depends on many factors, most important of which is the nature of the organism. From the standpoint of the irradiation of foods, the following kinds of organisms are important:

Viruses and Rickettsia	Moulds
Bacteria	Insects
Yeasts	Helminths

5.2.1. Viruses

Viruses are the smallest living entities. They are dissimilar from true organisms in that they have no respiration and are dependent on a host for food and enzymes. They are capable of reproduction and can affect their host. They can infect both plants (including bacteria) and animals. The principal concern with respect to irradiation of foods is to inactivate viruses present which have a health or economic significance. Heating is a very effective agent for inactivating viruses, so that foods which are normally cooked either during processing or in preparation for the table are not usually cause for concern. The only exception is the chance of post-preparation contamination of a food by a food-handler.

Foods have been incriminated in the transmission to humans of the viruses of poliomyelitis and infectious hepatitis. These viruses may contaminate a food either as a consequence of a human carrier handling the food or, especially in the case of shell-fish, owing to exposure to polluted waters. There is also the possibility of contamination of field-crop foods, such as vegetables, through application of human waste or sewage derivatives (e.g. sludge from waste treatment) as fertilizers.

A major problem, largely of economic significance, is the contamination of raw meat with the virus of foot-and-mouth disease. This virus does not usually infect humans but does attack many animals, including the usual domesticated meat animals. The disease exists in widespread areas of the world, only North America, Australia and New Zealand being free of it. Because of the dangers of transmitting the disease to the animals of the importing country, the disease-free countries prohibit the import of raw meat from those areas in which the disease exists. Other similar animal viral diseases include rinderpest and swine fever.

Radiation can inactivate viruses only at high doses. Thirty kGy (3 Mrad) have been shown to inactivate foot-and-mouth disease virus suspended in an aqueous medium. For inactivation in the dry state, 40 kGy (4 Mrad) were required.

Most of the doses demonstrated to be effective in raw products would produce undesirable effects on the food. There is little difficulty with products that have been heated to 60–70°C for a short time, as such temperatures inactivate the virus.

5.2.2. Bacteria

Bacteria are generally present in all foods except those processed to destroy a natural contamination. Control of food spoilage usually involves control of bacterial action. Certain bacteria, which can be carried by foods, are pathogenic to man. The nature of the food, its treatment and the storage conditions affect the bacterial pattern.

There are a large number of kinds of bacteria differing in morphological and physiological characteristics. An important classification basis is grouping as spore-formers or non-spore-formers. Both groups can exist as vegetative forms, but certain of them undergo change in the proper conditions to form spores. Spores are more resistant to stress conditions than vegetative cells. Under proper conditions, spores can germinate and grow in the vegetative form.

In the case of foods, concern about bacteria involves the following:

(a) Bacteria can cause sensory and other changes frequently considered undesirable and usually associated with spoilage;

(b) Certain organisms growing in a food produce a toxin harmful to man;

(c) Certain organisms in food can infect man and animals and thereby cause a disease condition.

Many food-preservation measures are directed towards control or destruction of spoilage microorganisms. Radiation is capable of destroying microorganisms and consequently is an applicable preservation agent. As would be anticipated, radiation acts through direct and indirect action, since water is generally a component. The action of radiation on bacteria is influenced by the following: amount of radiation; species and strain of bacteria; concentration or number of bacteria; chemical composition of medium; physical state of medium; and post-irradiation storage conditions.

The sensitivity of an organism to radiation is conveniently expressed in terms of the number of grays (Gy) required to accomplish the kill of a fraction of the population. This is done rather than attempting to determine the amount of radiation required to kill 100% of the population. Not only would it be difficult to do this experimentally, but the amount of radiation required is dependent on the number of organisms in the population. Therefore, a procedure is used that kills 90% of the population present. The result is expressed as the D_{10} value, or the treatment required to reduce the population by a factor of 10.

In the case of treatment with ionizing radiation, if

N_0 = initial population
N = population after dose D
D = rads
D_{10} = rads to reduce population by a factor of 10 (10% survival), then

$$\log_{10} \frac{N}{N_0} = -\frac{1}{D_{10}} D$$

D_{10} is commonly called the decimal reduction dose, or the D_{10} value. D_{10} values for a number of organisms commonly associated with food are given in Table IX to give an indication of the differences due to species and spore formation.

Figure 13 shows a plot relating dose to surviving fraction of microorganisms. The $\log_{10}(N/N_0)$ plotted against the dose is a straight-line relationship, provided

TABLE IX. D_{10} VALUES OF SELECTED BACTERIA

Bacterium	Irradiation medium	D_{10} (Gy)[a]
Ps. aeruginosa	Nutrient broth	30
Ps. fluorescens	Nutrient broth	20
Ps. geniculata	Nutrient broth	50
E. coli	Nutrient broth	100–200
Staph. aureus	Nutrient broth	100
Staph. aureus .	Dry	650
Salm. senftenberg	Meat-and-bone meal	500
	Liquid whole egg	170
	Dried egg	450–600
Salm. typhimurium	Meat-and-bone meal	600
Micrococcus radiodurans	Raw beef	2500
	Raw fish	3390
Moraxella osloensis	Raw beef	4770–10 000
Acinetobacter calcoaceticus	Raw beef	4050–8140
B. subtilis (spores)	Saline	2600
B. subtilis (spores)	Pea purée	350
C. botulinum (spores)		
Type A 12885	Phosphate buffer	2410
	Canned chicken	3110
	Canned bacon	1890
Type A 33	Cooked beef	3900
Type B 53	Phosphate buffer	3290
	Canned chicken	3690
	Canned bacon	2040
C. sporogenes (PA 3679/52)	Phosphate buffer	2090
C. perfringens	Aqueous suspension	1200–2000

[a] To convert to krad, divide by 10.

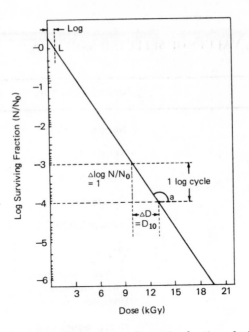

FIG.13. *Relationship between radiation dose and surviving fraction of microorganisms.*
(To convert kGy to krad, multiply by 100.)

the radiation effect is the direct effect and one ionization causes a kill. Such an effect would be independent of dose rate, temperature, concentration of micro-organisms, and the physical state and composition of the medium. The efficiency of a given type of radiation would depend on the target size and would relate to the number of ionizations produced per unit volume.

The target theory explains to a considerable degree the action of radiation on bacteria. Deviations from the straight-line relationship mentioned above occur, indicating that some additional mechanisms are involved. Figure 14 shows the effects of temperature (including physical state) and concentration of organisms upon the D_{10} value for Type-C *Clostridium botulinum* spores. In this case the D_{10} value increases as the temperature is lowered. These data also demonstrate that the D_{10} value varies with concentration of spores. Figure 15 shows the effect of a varying composition of medium on the surviving fraction of *Escherichia coli*.

Within wide limits, dose-rate variation appears not to affect D_{10} values. Some types of ionizing radiation are more effective than others in killing bacteria. These effects appear to be related to the density of ionization produced or, perhaps more precisely, to the energy transferred to the absorber from the radiation. Gamma rays, X-rays and electron beams, however, appear to be about equivalent in their action on bacteria.

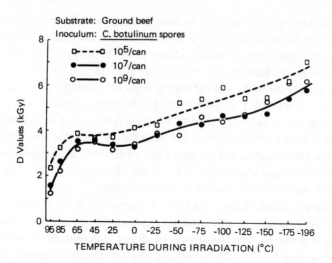

FIG.14. *Effect of temperature during irradiation on* D_{10} *values of spores of* Clostridium botulinum. *(To convert kGy to Mrad, divide by 10.)*

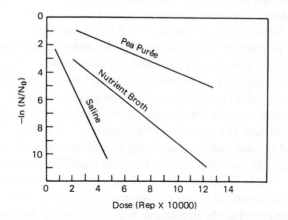

FIG.15. *Effects of γ-rays on* Escherichia coli *suspended in three different media. (To convert Rep to Gy, multiply by 0.093.)*

It should be recognized that while departures exist from the simple relationship between dose and kill predicted by the target theory, the specific effect cannot be predicted. The effect of variables such as medium, composition, physical state, etc., on each organism must be determined.

More than one consideration applies to the irradiation of foods for the purpose of affecting bacteria present. One objective may be to reduce the number of bacteria present in order to limit spoilage action, since bacterial spoilage is normally associated with growth of the contaminating bacteria. As the numbers increase, marked quality changes may occur in the food, and one way to prevent these changes is to hold down the level of bacterial populations. This can be accomplished by irradiating the food sufficiently to reduce the population to a level such that the time to reach an undesired value is extended. In this way spoilage is delayed. Usually the food is subjected to otherwise normal handling. It is clear that this kind of use of radiation does not seek initially to destroy all bacteria present, but permits some to survive. This use of radiation has been referred to as *radurization,* which parallels one use of the term *pasteurization* used in connection with heat treatment.

Radurization is not always a process of simple reduction of a bacterial population. As indicated above, different organisms have different sensitivities to radiation, and the kill of a population of mixed organisms is not uniform. The irradiation may lead to a changed flora and a different outgrowth pattern. Such changes may affect the ultimate spoilage of the food and require evaluation to protect consumers against unusual health hazards, or to avoid misleading them by causing a spoilage condition which they do not recognize.

There is also the possibility of mutations of the organisms surviving the radiation treatment. Mutations do occur, but no problems with mutants have been observed. Some microorganisms can build up a resistance to radiation but, here too, in the practice of the irradiation process this phenomenon presents no problem. Repeated exposure of a particular organism or its own progeny is not probable in irradiating foods. Irradiation is a 'once-through' process.

Radiation may also be used to control a specific organism whose presence in a food may be objectionable, for example, because it constitutes a health hazard. An example of this would be presence of Salmonellae. The radiation treatment may be designed to relate solely or primarily to such an organism, and the dose level to obtain an adequate reduction in numbers would be governed primarily by the D_{10} value of the organism. The term *radicidation* has been suggested for treatment by radiation to eliminate non-spore-forming pathogens.

Finally, the object of the irradiation may be to destroy all spoilage micro-organisms present; this may be to obtain a sterile product which, with suitable packing, will have an indefinite storage life at temperatures above those of refrigerator storage. In this case, consideration needs to be given to (a) elimination of all organisms whose growth would cause product spoilage, and/or (b) elimination of all organisms that might cause a consumer health hazard. The term *radappertization* has been suggested for this type of processing using ionizing radiation.

There is no absolute value for the amount of radiation required to destroy all microorganisms present. The destruction of microorganisms follows a statistical pattern, and for sterile products the objective is to reduce the probability of survival to less than one organism in an initial population of 10^{12} organisms. In attaining this objective, consideration is given to:

(a) The specific type of significant organisms;

(b) The initial inoculum level of these organisms; and

(c) The D_{10} value of the organisms in relation to the D_{10} value of other organisms likely to be present.

For non-acid, low-salt foods, the organisms of concern are the different types of *Clostridium botulinum.*

This spore-forming species can be present in foods and is hazardous to man because of toxin production. The D_{10} values for the various types of *Cl. botulinum* are somewhat different, but are all high (see Table IX). This is the most radiation-resistant group of organisms likely to be encountered in foods. There are several bacteria, such as *Micrococcus radiodurans, Moraxella osloensis* and *Acinetobacter calcoaceticus,* which are more radiation-resistant than *Cl. botulinum* (see Table IX). All these bacteria are, however, quite heat-sensitive and are easily destroyed by the heating given radappertized products for enzyme inactivation. Consequently they can be ignored in determining the dose requirements for sterile products. Irradiation of foods to obtain sterility has been gauged in terms of the D_{10} value for *Cl. botulinum.* All other organisms are presumed to be destroyed by setting the process for killing *Cl. botulinum.*

The level of contamination of a food with the spores of *Cl. botulinum* is the remaining factor for dose determination. The standpoint has been taken that the same standards which have been used in setting process schedules for heat sterilization of foods should be used for irradiation. The heat process schedules assume an initial population of 10^{11} spores per unit or package. Hence, to reach a population of less than one, the reduction accomplished by heat is 12 log cycles. To accomplish this with radiation requires 12 times the D_{10} value determined for *Cl. botulinum,* which is roughly 45 kGy (4.5 Mrad).

The D_{10} value, however, is affected by the environment in which the micro-organism exists (Figs 14 and 15). It is therefore necessary to use a value exactly applicable to the particular radappertized product and to the precise irradiation conditions. In practice, a 'minimum radiation dose' (MRD) for each product is determined from inoculated pack studies. The inoculum used consists of 10^6 spores per unit of each of ten different strains of Type-A and Type-B *Cl. botulinum.* One hundred replicates are used and a range of doses applied at the designated process temperature. Sterility is gauged on the basis of the absence of recoverable viable *Cl. botulinum* organisms after six months of incubation.

The 12-D requirement applies to low-acid (pH greater than 4.5), low-salt, high-moisture foods. Foods having chemical compositions that are different may require less radiation since *Cl. botulinum* may not grow and produce toxin on such foods. The principal argument against the use of the 12-D dose is the unlikelihood of an initial population of 10^{11} spores per unit ever being encountered. Arguments supporting the 12-D concept are the long-term success record of the heat process for canned foods, which uses this standard, and the viewpoint that statistically it makes no difference whether 10^{11} organisms exist in one unit (e.g. one can of product) or are distributed over a larger number of units. It is because of these latter considerations that the 12-D requirement has been accepted for radappertized low-acid, low-salt, high-moisture foods.

5.2.3. Yeasts and moulds

Yeasts and moulds are frequently present in foods, and their outgrowth can produce changes that are undesirable and cause spoilage. On nuts and grains in particular, some moulds, e.g. *Aspergillus flavus,* produce mycotoxins such as aflatoxin, which is a carcinogen. Certain species of both yeasts and moulds serve useful purposes in the production of foods such as cheese.

Yeasts and moulds have a sensitivity to radiation comparable with some non-spore-forming bacteria. There is a substantial variation according to species. The lethal dose for yeasts is approximately in the range 4.65 to 20 kGy (465 krad to 2 Mrad) and that for moulds ranges from 2.5 to 6.0 kGy (250 to 600 krad).

Considerable work has been done on the control of moulds which cause rotting and softening of certain fruits. In most cases the dose to kill the moulds is higher than the fruit can tolerate; the tissue becomes soft owing to degradation of pectic substances of the fruit. This softening can make the fruit susceptible to handling damage and rotting. Yeasts seldom cause fruit spoilage, but they do spoil fruit juices and other fruit products. Here, too, the dose requirement to effect preservation may be too high in that flavour changes occur.

One effective way to reduce the dose requirement for yeast and mould control is to use both radiation and mild heat.

5.2.4. Insects

Foods contaminated with insects are generally regarded as unfit for human consumption. In some cases the transport of insect-infested foods from one area to another results in distributing the insect. If the insect is harmful, say to certain crops, such food movement may be prohibited or the product may be subject to control and application of some processes, such as fumigation, to destroy the infestation.

Radiation can be a useful agent for controlling insect infestation of foods. We shall consider only the application of radiation to foods and not the use of radiation for sterilization of insects, per se, as a general population control measure. Insects are relatively insensitive to radiation. As with other organisms, the effects of radiation on insects are closely related to the effects on constituent cells. For cells, the sensitivity to radiation is in direct proportion to their reproductive activity and inversely proportional to the degree of differentiation. During the larvae period of insects very little cell division occurs. Cell division and differentiation of tissues occur during embryonic development in the egg, and there are brief periods just before moulting (if this process is normal) and in later stages of pupation. Dividing insect cells are as sensitive to radiation as are cells of vertebrates, but the generally static life of adult insects makes them insensitive to radiation. Certain cells of the adult, however, maintain a metabolic activity, namely the cells of the gonads, and these are sensitive to radiation. Hence relatively low doses cause sterilization or cause genetically deranged gametes. Higher doses are required for a lethal effect.

The general deteriorative effects of radiation on insects are: lethality, 'knockdown' (apparent lethality, followed by recovery), reduced longevity, delayed moulting, sterility, reduction of egg hatch, delay of development, reduction of food consumption, and respiration inhibition. These effects occur at certain dose levels. Reverse effects at other (lower) dose levels have been observed, including increased longevity, increased egg-laying, increase of egg-hatch, and stimulation of respiration.

The minimum dose levels of γ-radiation to arrest development of certain insects and a mite are shown in Table X. Similar data, to produce sterility in adults, are shown in Table XI. These show that moths are more resistant to the effects of radiation than are weevils. It must be pointed out that, at life stages below the adult, radiation produces sterility in these insect stages. Thus, if a larva develops from an irradiated egg it will not develop to the pupa stage. If an adult develops from an irradiated pupa, the adult will be sterile. Tables X and XI show that the following dose ranges apply: 130 to 250 Gy (13 to 25 krad) permit some development of eggs and larvae but prevent development to the adult stage. Four hundred to 1000 Gy (40 to 100 krad) prevent all eggs, larvae and pupae from developing to the next stage. Beetles in the adult stage require 130 to 250 Gy (13 to 25 krad) to produce sterility, whereas moths require 450 to 1000 Gy (45 to 100 krad). The mite tested requires 250 to 450 Gy (25 to 45 krad).

Control of insect infestations in foods may be considered in the following general terms. For immediate lethality, doses in the range of 3 to 5 kGy (300 to 500 krad) would be required. A dose of 1 kGy (100 krad) would probably be sufficient if lethality within a few days is the goal; a dose of 500 Gy (50 krad) would be sufficient if the goal was lethality within a few weeks and sterility of the living insects. From a practical point of view, the dose selection can be made on the basis of the tolerance that could be allowed. For stored products, a dose of 500 Gy (50 krad) will control the most resistant beetle species and the immature stages of moths. Any progeny of moths would be sterile owing to genetic damage. In any food, the most resistant organism likely to be encountered is the one which sets the minimum dose requirement. For uses such as quarantine control, a single insect may be all that is involved and, consequently, a dose specific for that organism is needed.

53

TABLE X. MINIMUM DOSAGE OF γ-RADIATION COMPLETELY ARRESTING DEVELOPMENT OF STORED-PRODUCT INSECTS AND A MITE

Organism	Stage exposed	Stage observed for effect	Dosage (Gy)[e]				
			132	175	250	400	1000
LEPIDOPTERA							
Plodia	Egg	Larva				X- - - - - - X	
interpunctella	Egg	Adult	X- - - - - -X				
	Larva	Pupa					>X
	Larva	Adult	X				
	Pupa	Adult					>X
Sitotroga cerealella	Egg	Larva					>X
	Egg	Adult	X- - - - - -X				
	Larva	Adult	X- - - - - -X				
	Pupa	Adult					>X
COLEOPTERA							
Tribolium	Egg	Larva					>X
confusum	Larva	Larva[a]	X				
	Larva	Pupa					>X
	Larva	Adult	X[b]				
	Pupa	Adult					>X
Lasioderma	Egg	Larva					>X
serricorne	Larva	Larva[a]				X- - - - - - X	
	Larva	Pupa					>X
	Larva	Adult	X[c]				
	Pupa	Adult					>X
Rhyzopertha	Larva	Pupa					>X
dominica	Larva	Adult		X- - - - - - X[d]			
	Pupa	Adult					>X
Attagenus piceus	Egg	Larva					>X
	Larva	Larva[a]			X- - - - - -X		
	Larva	Pupa	X				
	Larva	Adult	X				
	Pupa	Adult				X- - - - - - X	
Trogoderma	Egg	Larva				X- - - - - - X	
glabrum	Larva	Larva[a]		X- - - - - - X			
	Larva	Pupa	X				
	Larva	Adult	X				
	Pupa	Adult				X- - - - - - X	
ACARINA							
Acarus siro	Egg	Larva					>X
	Larva	Adult		X- - - - - - X			
	Hypopus	Adult				X- - - - - - X	

[a] Based on data obtained 21 days after exposure.
[b] One survivor after exposure to 450 Gy.
[c] One survivor after exposure to 1 kGy.
[d] One survivor after exposure to 450 Gy.
[e] To convert to krad, divide by 10.

TABLE XI. MINIMUM DOSAGE OF γ-RADIATION PRODUCING STERILITY
IN 100% EXPOSED ADULT STORED-PRODUCT INSECTS AND A MITE

Organism	Sex exposed	Dosage (Gy)[a]				
		132	175	250	450	1000
LEPIDOPTERA						
Plodia interpunctella	Male					>X
	Female					>X
Sitotroga cerealella	Male					>X
	Female				X- - - - - - - X	
COLEOPTERA						
Tribolium confusum	Male and female	X- - - - - - - X				
Lasioderma serricorne	Male and female		X- - - - - - - X			
Attagenus piceus	Male	X- - - - - - - X				
	Female	X- - - - - - - X				
Trogoderma glabrum	Male		X- - - - - - - X			
	Female	X				
ACARINA						
Acarus siro	Male and female			X- - - - - - - X		

[a] To convert to krad, divide by 10.

Radiation given in a single dose is more effective than if given in increments. Raising the temperature prior to irradiation sensitizes certain insects to radiation. Reduction of the atmospheric oxygen pressure increases resistance. Multiple exposures at substerilizing levels do not result in a build-up of radiation resistance.

5.2.5. Animal parasites

This group of organisms includes parasitic worms that can infest certain foods. Some of these can infest man and are therefore a matter of concern when present

TABLE XII. FOODBORNE HELMINTHS

Helminth	Carrier food	Human disease
Trichinella spiralis	Pork	Trichinosis
Taenia solium	Pork	Pork tapeworm
Taenia saginata	Beef	Beef tapeworm
Clonorchis sinenis	Fish	Liver fluke
Diphyllobothrium latum	Fish	Fish tapeworm
Paragonimus westermani	Crayfish	Lung fluke
Ascaris lumbricoides	Raw vegetable	Intestinal worms
Fasciolapsis buski	Raw fruit, water plants	Intestinal fluke
Fasciola hepatica	Watercress, lettuce	Liver fluke
Anisakis marina	Herring	Intestinal worms

in foods. Irradiation of the infested food is a possible method of controlling these organisms.

The principal helminths of interest in connection with foods are listed in Table XII. These organisms exhibit several forms during their life cycle but normally are in only one form as a food contaminant. Radiation is effective regardless of the form. For larval forms, with increasing dosage the effects are: sterility of the adult females; inhibition of normal maturation and localization; death. The requirement for sterilization of *Trichinella spiralis* is about 120 Gy (12 krad); for inhibition of maturation about 200 to 300 Gy (20 to 30 krad); and for death about 7.5 kGy (750 krad). For devitalization of the tapeworm of beef, 3 to 5 kGy (300 to 500 krad) may be necessary. A dose in excess of 6 kGy is needed to inactivate *Anisakis marina* in herring.

5.2.6. Plants

A great deal of information is available on the effects of radiation on living plants. Interest in this area in connection with the irradiation of foods is necessarily limited, and effects on growth and reproduction are outside this interest. Of concern, however, are those foods which exhibit certain life processes during the period between harvest and consumption, i.e. many raw fruits and vegetables.

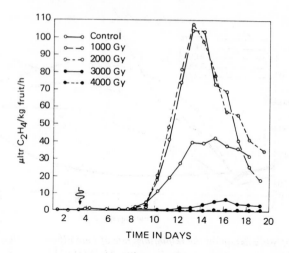

FIG.16.   *Effect of γ-irradiation on ethylene production by Bartlett pears. (To convert Gy to krad, divide by 10.) Data from Adv. Food Res.* **15** *(1966) 110.*

5.2.6.1. Fruits

Fruits may be classified, according to their respiratory behaviour during ripening, as of either the climacteric or non-climacteric class. Climacteric fruits exhibit a slowly declining respiration rate which reaches a minimum just before the onset of ripening. As ripening begins, respiration increases greatly and reaches a peak as the fruit becomes ripe. Final degradation of the fruit (senescence) is accompanied by a declining rate of respiration. Non-climacteric fruits are often fully ripe at harvest and show a slowly declining rate of respiration without any period of peak activity.

For climacteric fruits the pre-climacteric respiration minimum is a key point with regard to response to stimuli, including radiation. Irradiation of fruits in the pre-climacteric stage produces a greater response to radiation than fruits treated beyond the onset of the climacteric.

Figure 16 presents data showing the effect of γ-radiation on ethylene production in the pre-climacteric Bartlett pears. Pears irradiated at the climacteric peak show a decrease in ethylene production. For relatively high doses (3 to 4 kGy (300 to 400 krad)) of irradiation of pears in the pre-climacteric stage, ripening is inhibited and cannot subsequently be brought about by exposure to ethylene.

Not all fruits behave like pears. Peaches and nectarines, for example, when irradiated with doses as high as 6 kGy (600 krad) are stimulated to ripen. Figure 17 shows data on the changes in respiration rate of Late Elberta peaches caused by irradiation. Figure 18 shows the changes in rate of ethylene production for this fruit. Unlike pears, the peaches ripen with high doses and remain sensitive to ethylene.

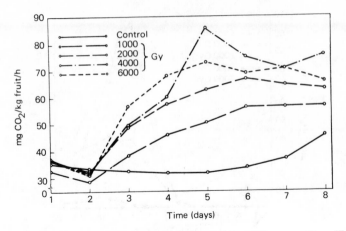

FIG.17. *Effect of γ-irradiation on the respiratory rate of Late Elberta peaches. (To convert Gy to krad, divide by 10.) Data from Adv. Food Res.* **15** *(1966) 112.*

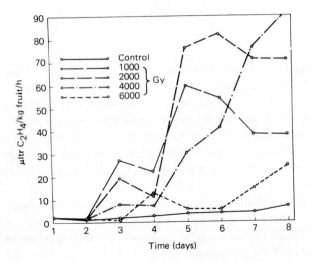

FIG.18. *Effect of γ-irradiation on rate of ethylene production by Late Elberta peaches. (To convert Gy to krad, divide by 10.) Data from Adv. Food Res.* **15** *(1966) 113.*

Since the response of climacteric fruits to radiation is related to the position in the climacteric sequence, the use of radiation in irradiating fruits for preservation or other purposes must take this into account in order to minimize adverse effects and secure desirable ones.

Non-climacteric fruits show a response to radiation somewhat similar to climacteric fruits. Respiration rate and ethylene production are increased. Since

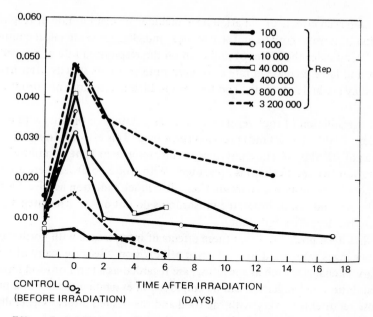

FIG.19. Effect of γ-irradiation on the oxygen uptake of potato tubers. (To convert Rep to Gy, multiply by 0.0093.) From SUSSMAN, A.S., J. Cell. Comp. Physiol. **42** (1953) 272.

ripening is not involved, these effects do not indicate climacteric induction.

Varietal differences can be important in the response of a fruit to radiation. The Gros michel banana, for example, undergoes delayed ripening upon irradiation, whereas the Basri variety does not. The skins of certain bananas become brown. This is ascribed to an increase in polyphenol oxidase activity in the skin and pulp, possibly due to radiation-induced cell damage.

Radiation can produce changes in chemical composition of fruits. Such changes include: destruction of ascorbic acid; conversion of protopectin to pectin and pectate; degradation of cellulose and starch; destruction of certain acids such as malic (in apples); and pigment changes. Texture changes appear to be associated with pectin changes and can be a limiting factor in the amount of radiation that can be employed. Softening can be reversed by treating some fruit with calcium salts. Reversal of the softening of strawberries occurs spontaneously on storage after irradiation.

5.2.6.2. Vegetables

Raw vegetables, like raw fruits, consist of slowly metabolizing tissue. Radiation can affect the rate of this metabolism, the specific effect being related to the radiation dose. Most studies have been with doses in the 10 Gy (1 krad) range. Observed effects of radiation have included changes in rate of respiration, inhibition of normal growth and senescence, and changes in chemical composition. Figure 19 indicates the effect of radiation on the respiration rate of potato tubers. A quick and large increase in rate of oxygen uptake occurs shortly after irradiation, followed by gradual reduction. Too low or too high a dose does not produce this effect.

The irradiation of root vegetables, such as potatoes or onions, with doses of the order of 100 Gy (10 krad) prevents sprouting. This effect is irreversible. Greening of the skin of white potatoes in the presence of light is inhibited. Suberization (wound healing) is prevented. With white potatoes there are varietal differences in the response to irradiation. Too much radiation accelerates rotting.

The cap and veil of irradiated mushrooms do not open. Irradiated asparagus spears do not lengthen with time.

It has been postulated that these effects of interference with normal post-harvest changes are related to suppressed cell division. Because many of these post-harvest changes, such as sprouting, are regarded as a form of food losses, their inhibition by irradiation can be employed in extending the life of a product. The dose requirements vary with the food and the desired effect. Generally they are among the smallest used in food irradiation.

6. PRESERVATION OF FOODS

The main reasons for preserving foods are to make them available after production or harvest or at places different from where they were produced or harvested. As discussed in earlier sections of this Manual, there are a number of ways in which foods spoil or become unsuitable for consumption. Likewise, a variety of preservation methods to prevent spoilage are in use. Assessment of spoilage and preservation requires definition of quality factors that describe suitable and unsuitable foods.

6.1. Quality factors important to foods

6.1.1. Nutritive value

The principal function of foods is to nourish and sustain the life of the consumer. Hence foods must be the source of the nutrients required for these functions. A major consideration is that the foods provide the energy needed for life, and so the caloric content of a food is a major quality attribute. The macro-constituents of food (other than water) are fats, carbohydrates and protein. These can all be utilized by the consumer to supply energy. The approximate caloric content per gram of these constituents as determined by heats of combustion of digested nutrients is as follows:

Fat	9 cal
Carbohydrate	4 cal
Protein	4 cal

The physiological energy obtained by the consumer differs somewhat from these values, depending on the type of food and the individual food. Typical compositions of selected foods are shown in Table XIII.

The physiological energy requirement of a human is a function of age, sex, size, environment and activity of the individual. Table XIV gives estimates of caloric requirements of adults according to age.

Fats, carbohydrates and proteins, while sources of energy, also have other functions in the work of the body. Fats, in addition to providing energy, are carriers of micronutrients such as the fat-soluble vitamins. Proteins are sources of amino acids. Certain amino acids are regarded as 'essential' in human nutrition in that they must be present in the consumed food as they cannot be synthesized by the human. Other amino acids which occur in many foods are not in this essential category but are utilized by the body. Proteins, and foods containing them, can therefore be rated according to their amino acid composition, both as to which amino acids are present and the amounts of each. The need for particular amino acids varies with age, children having special requirements. In general, proteins of animal origin are of higher nutritional quality in terms of an index of essential amino-acid composition than are proteins of plant origin. Proteins of plant origin may be lacking in one or more essential amino acids. Proteins exist in every cell and are essential constituents of body tissues.

Vitamins and minerals are microconstituents of foods that play key and complex roles in the body functions. Vitamins may be classified as fat-soluble or as water-soluble. Generally, vitamins used in the body functions are obtained from the ingested foods. Minerals have no other sources than food and water taken into the body. Hence, the vitamin and mineral contents of foods represent important quality indexes. Preservation methods or improper storage can reduce the vitamin content of a food and in this way degrade its nutritional value.

It is recognized that good nutrition requires a proper balance of nutrients and, while much in the area of nutrition has been learned, there is still insufficient knowledge to specify with precision the optimum levels for many nutrients. The determination of such levels is complicated by variations in needs with age, sex, size, environment and activity of individuals. For many reasons, good nutrition is usually considered to result from a diet of a variety of foods, provided such a diet includes essential nutrients in appropriate amounts.

TABLE XIII. COMPOSITION OF SELECTED FOODS — EDIBLE PORTIONS

Food	Per cent fat	Per cent carbo- hydrates	Per cent protein
Apple, raw, pared	0.3	14.1	0.2
Banana, raw, common	0.2	22.2	1.1
Beans, raw, white	1.6	61.3	22.3
Beef, raw, round	12.3	0.0	20.2
Cabbage, raw, common	0.2	5.4	1.3
Cauliflower, raw	0.2	5.2	2.7
Cheese, cheddar	32.2	2.1	25.0
Chicken, raw, skinned			
Light meat	4.9	0.0	18.6
Cod, raw	0.3	0.0	17.6
Eggs, chicken, raw, whole	11.5	0.9	12.9
Haddock, raw	0.1	0.0	18.3
Lamb, leg, raw	21.0	0.0	16.9
Milk, cow, whole	3.7	4.9	3.5
Orange, raw, peeled	0.2	12.2	1.0
Pork, loin, raw	28.0	0.0	16.4
Potatoes, white, raw	0.1	17.1	2.1
Rice, white, polished, raw	0.4	80.4	6.7
Shrimp, raw	0.8	1.5	18.1
Soy beans, mature, raw	17.7	33.5	34.1
Sweet potato, raw	0.4	26.3	1.7
Tuna, blue fin, raw	4.1	0.0	25.2
Veal, loin, raw	15.0	0.0	18.6
Wheat flour, bread	1.1	74.7	11.8

6.1.2. Safety

Not only must a food have nutritive value and other desirable characteristics; it must not contain anything harmful to the consumer. No food, however, can be made completely safe. It can be safe only to the degree that the probability of a health hazard for the consumer is extremely small. For example, the radappertization process (and the comparable thermal process) is designed to reduce the probability of survival of a viable spore of the toxin-producing *Cl. bot.* to one

TABLE XIV. ESTIMATE OF CALORIC REQUIREMENTS OF ADULTS ACCORDING TO AGE
(calories per day; mean temperature 10°C)

Age	Men	Women
20 - 30	3200	2300
30 - 40	3104	2231
40 - 50	3008	2162
50 - 60	2768	1990
60 - 70	2528	1817
70	2208	1587

in 10^{12}. This risk, although small, must be considered in the light of the benefit to the consumer. The judgement upon which the risk is either accepted or rejected is based on the risk-benefit balance. In the case of radappertized foods, the benefit relates the availability of shelf-stable foods to their utility in the diet of the consumer. In consideration of this particular benefit, and considering only the botulinum hazard, which has a very low probability of occurrence, the risk-benefit balance clearly favours proceeding with radappertized foods.

Hazards associated with foods can be classified essentially as chemical or biological in nature. Some chemical substances are capable of causing impaired health, or of preventing normal life processes such as reproduction, or of causing death. Some may not harm the immediate consumers, but may affect their progeny. Hazards classified as biological generally result from the contamination of a food with an organism that can produce a disease condition in the consumer. Such organisms include bacteria, parasites (mainly helminths) and viruses.

Foodborne chemical toxicants occur in several ways. Some may be present naturally, i.e. they may be normal constituents, or they may have entered from the environment in which the food plant or animal grew. Others may have been added, whether intentionally or not. Substances known to be toxic are not usually added to foods intentionally. However, a few additives, e.g. sodium nitrite, can be toxic at certain levels of intake. Unintentional additives are either accidental contaminants or are present as the result of a treatment or handling procedure. For example, it is possible for a pesticide applied to a field crop to carry through to the eating table. Some chemical toxicants are the result of biological action on the food by certain bacteria or moulds. They produce potent bacterial toxins or mycotoxins.

There is a need for a system of quality assessment for detecting chemical and biological hazards in foods. Good practices designed and carried out to avoid hazards throughout the food chain are essential and are often imposed by govern-ment regulation to protect the consumer. New knowledge, however, can reveal hitherto unsuspected hazards and call for changes in existing practices.

Because radiation can cause chemical changes in foods, and because of certain potential biological hazards, the assessment of the safety of irradiated foods for human consumption has been considered necessary. Similar concern exists for some other food treatments. In recent years, the methods of safety assess-ment have been refined and extended with the result that the magnitude of the task has increased. On the other hand, a better understanding of the action of radiation on foods has also developed, allowing the scope of the evaluation work to be narrowed to particular areas of concern.

6.1.3. Sensory acceptability and its testing, market testing and functionality

Sensory characteristics include those quality aspects of a food which, one way or another, are discernible by the senses of the consumer. These characteristics are generally used in his acceptance and selection of food, and govern to a considerable degree the pleasure and satisfaction he derives from its consumption. Sensory characteristics include: colour, odour, taste, texture, shape, size, tenderness, viscosity, uniformity, non-uniformity, temperature, etc. These are subjective characteristics, and their meaning to an individual is based on many human factors such as conditioning, experience, custom, social status, as well as apparently inherent preferences. They are important characteristics because in many circumstances they govern the decision on what is purchased and eaten. Regardless of nutritional value, variations in these sensory characteristics frequently affect the market value of a food, since consumer demand will vary according to preference.

Efforts have been made to devise physical or chemical methods of measuring these sensory characteristics. Such approaches have met with varying success; some are used regularly in research and product evaluation or in process control. In many cases, however, the subjective reaction of the human senses to a food characteristic is too complex to be obtained in such an objective way, and the only really dependable methods employ human observers who use their sensory capabilities in a reasonably precise fashion to evaluate a food. This type of evaluation has been highly developed. It usually involves 'panel' testing, in which a group of individuals examine the food and each makes a judgement on a particular quality characteristic. The individual judgements are then brought together and the group judgement ascertained. In some circumstances the judge-ment of a single tester may be the only one obtained.

Panels of judges are of two principal types: (1) affective and (2) analytical. Affective panels are concerned with evaluation of preference and/or acceptance and/or opinions of a food. Analytical panels evaluate differences or similarity, and quality and/or quantity sensory characteristics. Analytical panels may be further classified as (a) discriminative and (b) descriptive. Discriminative panels measure whether or not samples are different (difference-similarity) or the ability of individuals to detect sensory characteristics (sensitivity). Descriptive panels measure qualitative or quantitative characteristics. Affective panels can offer an indication of acceptance of a food by the ultimate user. Analytical panels indicate similarities or differences without regard to consumer reaction.

Because of their great importance in guiding research and in the marketing of foods, a number of techniques have been devised for carrying out the purposes of the panel evaluations and, where needed, for qualifying panelists. Since consumer acceptance is a critical factor in irradiated foods, the use of appropriate panels and techniques for evaluation is very important in the total development of the process.

The particular method of evaluation used depends on the type of food involved and the type of information desired. Table XV lists commonly used methods, and Table XVI gives recommendations for their use.

In general, affective-type panelists are untrained and are randomly selected to represent the target population of consumers. Usually a minimum of 24 panelists is required, but larger numbers are more often employed. Analytical-type panelists must have demonstrated ability to determine differences and to reproduce their judgements. Normally they are screened for their interest and trained to function in the panel. They are periodically requalified. The analytical panel usually consists of a small number. Product variability and judgement reproducibility determine the number of panelists.

While the performance of panels is the key part of an evaluation, sample preparation and presentation and the physical conditions of the test are also very important. All these must be standardized in order to yield results which are valid on a comparative basis. Physical facilities for analytical panels and for some affective panels are usually specially designed to assist the panelists in making independent judgements.

The design of the evaluation study should be in accordance with the study objectives and of a nature to yield statistically significant conclusions. Various statistical treatments of the test data are available.

Table XVII gives an example of a scale which can be used with irradiated foods for threshold irradiated flavour determination and for other purposes such as new product development, product matching, product improvement or quality rating.

Table XVIII shows a common verbal hedonic scale which can be used for a variety of purposes mainly associated with consumer acceptance. The use of this scale in preference ratings is illustrated in Table XIX.

Related to evaluation of sensory acceptance is market testing, the object of which is to determine how acceptable a product is when available in a market

TABLE XV. SENSORY EVALUATION METHODS

1. Single sample
2. Paired comparison
3. Duo-trio
4. Triangle
5. Rank order
6. Rating difference (scalar difference from control)
7. Quality rating (scalar scoring)
8. Hedonic (verbal or facial)
9. Flavour profile
10. Texture profile
11. Threshold
12. Dilution
13. Food action scale
14. Magnitude estimation
15. Quantitative descriptive analysis

From Food Technol. **30** (1976) 43.

situation. It involves not only sensory acceptance but other factors as well such as cost, convenience, utility and competitive or alternative product availability. The test product may be offered in competition with similar products or, if the product is new, the test may be directed towards determining what market position it might secure. Sales promotion may be a part of the market test.

Market testing is normally a necessary step or set of steps carried out before true commercial marketing. The scale of the test, at least initially, may be relatively small: for example, distribution of the product to one store or to one city. Market testing involves a variety of techniques which are best carried out by market-testing specialists acquainted with local conditions.

An important quality aspect of certain foodstuffs concerns their ability to perform a particular function in the preparation of a compound or formulated food. Flour, for example, must be capable of producing a loaf of bread with suitable texture, and while this functional property is not apparent until the loaf is baked, it is inherent in its value as a flour. Various methods for determining the functionality of particular foods for certain uses are available.

6.1.4. Stability

The period of time that a food will keep in satisfactory condition is important. This period can be of indefinite or of limited duration, depending on how the

66

TABLE XVI. RECOMMENDED USES OF SENSORY EVALUATION METHODS

Type of problem	Appropriate test methods from list in Table XV
New product development (no prototype)	1, 2, 7, 8, 9, 10, 13, 14, 15
Product matching	2, 3, 4, 6, 7, 9, 10, 14, 15
Product improvement	2, 3, 4, 5, 6, 7, 9, 10, 13, 14, 15
Process improvement	2, 3, 4, 5, 6, 7, 8, 9, 10, 13, 14, 15
Cost reduction and/or selection of new source of supply:	
1. Maintain original specification	2, 3, 4, 6, 7, 9, 10, 14, 15
2. Create different product	1, 5, 7, 8, 9, 10, 13, 14, 15
Quality control	2, 3, 4, 6, 7, 9, 10, 12, 14, 15
Storage stability	7, 9, 10, 14, 15
Product grading or rating by sensory evaluation	7
Selection of best sample	2, 5, 7, 9, 10, 14, 15
Consumer acceptance and/or opinions	1, 2, 8, 13, 14
Consumer preference	2, 5, 14
Selection of panelists to be trained	3, 4, 7, 11, 12
Economic analysis	14
Correlating sensory with chemical and physical measurements	5, 6, 7, 8, 9, 10, 14, 15

From Food Technol. **30** (1976) 43.

TABLE XVII. SCALE OF IRRADIATION FLAVOUR INTENSITY

Numerical designation	Degree of flavour intensity
1	Imperceptible
2	Slightly perceptible
3	Perceptible
4	Slightly pronounced
5	Moderately pronounced
6	Pronounced
7	Very pronounced

TABLE XVIII. HEDONIC SCALE

Numerical designation	Verbal description
1	Like extremely
2	Like very much
3	Like moderately
4	Like slightly
5	Neither like nor dislike
6	Dislike slightly
7	Dislike moderately
8	Dislike very much
9	Dislike extremely

TABLE XIX. PREFERENCE TEST RATING SHEET

Sample No.					
	Like extremely	Like extremely	Like extremely	Like extremely	Like extremely
	Like very much	Like very much	Like very much	Like very much	Like very much
	Like moderately	Like moderately	Like moderately	Like moderately	Like moderately
	Like slightly	Like slightly	Like slightly	Like slightly	Like slightly
	Neither like nor dislike	Neither like nor dislike	Neither like nor dislike	Neither like nor dislike	Neither like nor dislike
	Dislike slightly	Dislike slightly	Dislike slightly	Dislike slightly	Dislike slightly
	Dislike moderately	Dislike moderately	Dislike moderately	Dislike moderately	Dislike moderately
	Dislike very much	Dislike very much	Dislike very much	Dislike very much	Dislike very much
	Dislike extremely	Dislike extremely	Dislike extremely	Dislike extremely	Dislike extremely

Please check for each sample the one set of words which best describe your reaction to it.

food is used. The choice of limited or indefinite preservation is related to the need or objective of storage and to what is attainable. Very often, the storage time requirement is only a matter of days; production, harvesting and processing can then be geared to the circumstance. On the other hand, certain foods, such as those produced seasonally, need to be stored from one harvest to the next or transported long distances, in which cases the preservation period may be many months or even longer.

It is essential to maintain stability of the food during the preservation period. Many of the quality indexes referred to above provide the means of judging the amount of change during storage and serve as guides in determining the amount of change that must be tolerated.

6.2. Spoilage agents for foods

Degradation of foods occurs in several ways, which may be classified as biological, chemical and physical. The particular spoilage pattern of a given food tends to be characteristic of that food under comparable preservation conditions. The condition of the product, type and extent of contamination, and the temperature of storage, are factors that markedly affect the rate of deterioration in quality.

6.2.1. Biological spoilage

A common kind of spoilage of a biological nature is associated with contamination of the food with organisms whose growth produces changes in the food which are regarded as undesirable. Bacteria, yeasts and moulds are the principal organisms that account for this type of spoilage. The food provides the nutrients for the growth of these microorganisms. Appropriate conditions of temperature and moisture are generally required in order to permit their growth. Foods tend to be contaminated with a characteristic flora, and the initial microbial population level is an important aspect of the spoilage pattern.

In a somewhat different manner, the contamination of a food with insects can 'spoil' the food. The spoilage in this case may be the objectionable presence of the insects themselves or it may be actual damage to the food.

Contamination of a food with pathogenic microorganisms may not produce spoilage of a sensory nature, but if such contamination constitutes a health hazard, this can be considered as a form of spoilage.

Some foods, such as fruits and vegetables, are themselves living organisms and, while they are stored, the normal life processes continue, leading to senescence, characteristic softening and deterioration. Ultimately, this development leads to changes which make the foods unacceptable.

6.2.2. Chemical spoilage

Chemical spoilage usually results from the interaction of food components or from reaction of the food with its environment. The reaction of sugars with proteins (Maillard or browning reaction) is an example of the former, in which this sugar-protein reaction produces undesirable flavour and colour changes. Rancidity development in a fat is an example of a food deteriorating through reaction with its environment (atmosphere oxygen). Chemical spoilage can be caused by active native enzymes present in the food.

6.2.3. Physical spoilage

Physical spoilage is perhaps the least important kind. It usually results from bruising, cutting or breaking a food during harvesting, transporting or handling. It can also result from loss of moisture, which changes the texture of the product, or from the uptake of moisture from the surroundings to a point where mould growth can take place and spoil the product.

6.3. Common food preservation methods

Most food preservation methods used today had their origins in prehistoric times. They have been refined and improved, elaborated, placed on a scientific basis, controlled and extended in application. The following is a list of these ancient methods:

(a) Drying, including dehydration
(b) Refrigeration (including freezing)
(c) Chemical preservation (e.g. salting)
(d) Fermentation
(e) Heat treatment (cooking, roasting, smoking)
(f) Packaging

To these prehistoric methods have been added three new processes:

(g) Canning
(h) Controlled atmosphere
(i) Irradiation

Canning was invented in the beginning of the nineteenth century. Controlled-atmosphere handling of foods, especially fruits and vegetables, is a very recent development. Irradiation is a basically new approach to food preservation.

The established preservation methods in all their multitudinous applications involve many steps, unit operations and the use of specially designed and built equipment. They accomplish the purposes for which they are carried out but cannot be regarded as without problems or without opportunities for improvement. Research to improve food preservation methods is an important activity in food science and technology today.

7. RADIATION PRESERVATION OF FOODS

7.1. General effects of radiation on foods

The effects of radiation on foods relate generally to effects on living organisms. With few exceptions, the radiation effects that cause changes in the intrinsic characteristics of foods are not of interest and may actually present problems. Useful general effects are of the following types:

(a) Control of microbial spoilage:
    (i) Radappertization
    (ii) Radurization

(b) Control of pathogenic non-spore-forming microorganisms (other than viruses)
(c) Control of foodborne pathogenic parasites
(d) Control of insects:
    (i) To prevent food damage or loss
    (ii) To prevent product contamination for aesthetic reasons
    (iii) To control distribution of insects when infested food is transported away from where it was produced

(e) Inhibition of sprouting and delay of senescence (of living foods)
(f) Quality improvement

The different kinds of application require different amounts of radiation. For convenience, they can be classified as (1) high dose (10 to 50 kGy (1–5 Mrad)), and (2) low dose (<10 kGy (<1 Mrad)). In general, high-dose treatments yield products that are sterile, or essentially so, and which when suitably packaged will keep indefinitely. The high dose is required not only to kill vegetative bacteria, moulds and yeast, but also spores. Viruses are not considered to be present in such applications owing to the high doses required. Enzymes, if present, are likewise not inactivated by the radiation but must be controlled by other means.

Low-dose treatments involve reduction of a microbial population, the control of organisms larger than bacteria, and the control of senescence of live foods.

Quality improvement applications involve changes in the normal intrinsic characteristics of a food, such as tenderness, and require specific doses, which may be either low or high.

7.2. High-dose irradiation of foods

7.2.1. Meats and poultry

The objective is to produce meat and poultry products that will keep for extended and essentially indefinite periods without refrigeration. The requirements for doing this are:

(a) The use of a radiation dose sufficient to destroy the most radiation-resistant microorganism associated with the product.

(b) Packaging of the product in a sealed container to prevent (i) microbial contamination after irradiation, (ii) chemical deterioration of the food by atmospheric oxygen, and (iii) moisture loss.

(c) Inactivation of enzymes indigenous to the products.

In setting up the irradiation process, the first requirement is to establish the dose needed. Reference to Table IX and to Section 5.2.2 will indicate that the most radiation-resistant organism of concern is the spore of *Cl. bot.*, Types A and B. Further, as stated in Section 5.2.2, radappertization uses the same standard as the thermal canning process: a 12-D reduction of spore count. Based on the procedures outlined in that section, minimum radiation doses (MRDs) have been determined for a number of meats and one seafood product (see Table XX).

Care must be exercised to deliver the MRD to all volume elements of the food. This requires selection of appropriate kind and energy of radiation. The penetration of 10-MeV electrons, for example, may not be sufficient for some product units owing to their thickness. When using γ-rays or X-rays, in order to ensure delivery of the MRD the manner in which the product is placed before the radiation source must be considered.

The importance of delivery of the MRD in the radappertization process differs from the importance of adequate heating in thermal canning. Whereas *Cl. bot.* is the most radiation-resistant organism of concern in radappertization, it is not the most heat-resistant. Thermal canning processes are therefore set to destroy microorganisms other than *Cl. bot.* which are heat-resistant. Their outgrowth in a product does not produce a health hazard as does *Cl. bot.*, but sensory spoilage may occur which, if identified, gives warning of inadequate processing and thus protects the consumer. No comparable safety factor exists with radappertized foods, and it is therefore mandatory that the delivery of the appropriate MRD be secured.

72

TABLE XX. MINIMUM RADIATION DOSE (MRD) FOR RADAPPERTIZED MEATS AND CODFISH CAKES

Food	Irrad. temp. (°C)	MRD[a] (kGy)
Bacon	5 to 25	25
Beef[b]	−30 ± 10	41
Ham[c]	5 to 25	31
Ham[d]	−30 ± 10	33
Pork	5 to 25	43
Codfish cakes	−30 ± 10	32
Corned beef	−30 ± 10	24
Pork sausage	−30 ± 10	27

[a] To convert to Mrad, divide by 10.
[b] With additives 0.75% NaCl, 0.375% Na tripolyphosphate.
[c] High $NaNO_2/NaNO_3$ (156/700 mg/kg) regular.
[d] Reduced $NaNO_2/NaNO_3$ (25/100 mg/kg).

The MRDs for the various products shown in Table XX reflect both the differences in product composition and the temperature of irradiation.

Before irradiation, the food is placed in the container and the container closed. Conventional metal ('tin') cans are satisfactory. A flexible pouch is also available (see Section 8.2).

Enzymes in meats are not inactivated completely by radappertization doses. Over a period of time they can cause undesirable product changes. Bitter flavours and crystals of tyrosine have been observed in stored radappertized raw meats. Heating the meat to 70–75°C inactivates the enzymes. In practice, the product is heated before it is placed in the container. This mild heat treatment for enzyme inactivation serves other purposes: it is sufficient to destroy the highly radiation-resistant microorganisms such as *Micrococcus radiodurans*. It also inactivates viruses and parasites.

The relatively high dose requirement for radappertized foods (the highest used in food irradiation) has caused a major problem with these products. If sufficient radiation is applied to any food, a characteristic foreign flavour and odour develop. This flavour is similar to that associated with heat scorching but not identical. Its intensity is dose-dependent. Different foods require varying amounts of radiation to produce a detectable 'irradiation flavour'. Pork, beef and

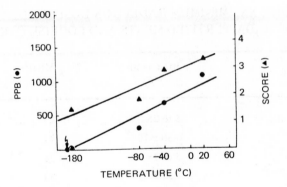

FIG.20. *Comparison of the relationship of component amount and flavour score of beef irradiated at 56 kGy (5.6 Mrad) as a function of temperature. An arrow denotes value for unirradiated control. Reprinted from J. Agric. Food Chem.* **23** *(1975) 1041.*

chicken irradiated at normal refrigeration temperatures develop a detectable flavour at about 2 kGy (200 krad). Radappertization doses applied at refrigeration temperatures therefore produce objectionable flavours in most meats and make them unacceptable to consumers.

Efforts have been made to identify the compounds responsible for this off-flavour. It has been reported that methional, 1-nonanal and phenyl-acetaldehyde are present in the ratio 20 to 2 to 1, respectively, and are the three substances formed on irradiation of beef which are the most important contributors to the characteristic irradiation odour. These three compounds are not, however, the only ones formed when meats are irradiated.

Direct bond cleavage of the various compounds making up meat accounts for many compounds that have been isolated. Lipids give rise to n-alkanes, n-alkenes and n-alkynes. Sterols yield normal and iso-alkanes. Proteins and peptides do not cleave at the peptide bond but at the side chains, giving rise again to hydrocarbons such as n-alkanes, benzene and toluene. Sulphur-containing proteins yield sulphides, disulphides and mercaptans.

Despite the knowledge of the compounds responsible for the undesirable flavour of irradiated meats, methods to prevent or suppress it have not been readily forthcoming. The best method devised so far has been irradiation at sub-freezing temperatures (−30 to −80°C). This suggests that the flavour compounds are formed not by direct action of the radiation but by indirect action of free radicals, probably originating in the water that is present. Lowering the temperature apparently eliminates the liquid phase, immobilizing the free radicals and preventing their interaction with the flavour constituents of the meat.

Figure 20 shows the variations with temperature of irradiation of flavour intensity and amounts of detected volatile radiolytic products of beef irradiated

TABLE XXI. EFFECT OF IRRADIATION TEMPERATURE ON PREFERENCE RATINGS* OF IRRADIATED HAM

Storage (months)	No. of taste-test panellists	35–44 kGy (3.5–4.4 Mrad)				Unirradiated control
		+5°C	−18°C	−40°C	−80°C	
1	30	−	5.9	5.9	6.8	7.5
1	30	5.6	6.1	6.4	7.1	6.9
4	30	5.5	5.8	5.6	6.6	6.1
12	32	5.4	−	−	6.2	6.9
12	32	6.1	−	−	6.8	6.4
Overall average		5.65	5.93	5.97	6.70	6.76

* Based on a 9-point hedonic scale; 9 is 'like extremely', 5 is 'neither like nor dislike', and 1 is 'dislike extremely'.

at 56 kGy (5.6 Mrad). It is clear that less irradiation flavour and smaller amounts of volatiles are produced at lower temperatures. The amount of irradiation flavour which may be tolerated has been studied with the objective of avoiding the use of very low temperatures. Table XXI shows the results of one study on ham. On the nine-point hedonic scale, a rating of 6 is considered satisfactory. For ham, at the dose level used, irradiation at −40°C is adequate to control the flavour change. Table XXII shows the acceptance of radappertized meats, chicken and two seafoods as determined by extensive consumer testing.

Since part of the bactericidal action of radiation is due to the indirect effect, it is reasonable to anticipate that irradiation at sub-freezing temperatures would affect this also. For the effect of temperature during irradiation on spores of *Cl. botulinum* see Fig. 14. Fortunately, it appears that only a small increase in dose is needed to kill *Cl. botulinum* compared to the large effect on flavour, making the overall value of low-temperature irradiation a gain. This points to the need for careful determination of the sterilizing dose for the product in question under the particular conditions of irradiation treatment.

Radappertization of meats is carried out anaerobically, i.e. the filled containers are vacuumized at the time of closing. Radappertized uncured meat from a freshly opened container displays an unusual pinkish or red colour. This changes quickly to the normal brown or grey of cooked uncured meat. Raw uncured meat, heated in the absence of oxygen, displays a similar colour that also changes on exposure

TABLE XXII. ACCEPTANCE OF RADAPPERTIZED MEATS, POULTRY AND SEAFOODS

Item	Irradiated		Unirradiated control	
	No. of evaluators	Rating[a]	No. of evaluators	Rating[a]
Ham	1 657	5.87	1437	6.66
Chicken	583	6.07	548	6.36
Pork	391	5.71	458	6.85
Beef	589	5.99	644	6.61
Bacon	25 656	6.16	—	—
Shrimp	539	6.09	849	6.43
Codfish cakes	531	5.40	578	6.30

[a] Based on a 9-point hedonic scale; 9 is 'like extremely', 5 is 'neither like nor dislike', and 1 is 'dislike extremely'.

to oxygen. The radiation-produced pink colour is apparently a reduced denatured myoglobin pigment which is easily oxidized. This phenomenon does not seem objectionable to consumers.

Anaerobic irradiation and post-irradiation holding prevent lipid oxidation.

7.2.2. Marine and fresh-water products

Many of the considerations relating to the high-dose irradiation of meats apply to marine and fresh-water products. In general, these are not as subject to flavour change as are meats, and therefore irradiated products such as shrimp, codfish (cakes) and lobster, with good sensory characteristics, have been studied.

7.2.3. Other foods

Bread has been reported acceptable when irradiated for the control of moulds with doses as high as 2 Mrad.

Milk and dairy products undergo severe flavour changes which make them unacceptable. Low doses on the surface of cheese have been successfully used to control surface mould growth.

Vegetables show varying responses to high-dose radiation treatments. Texture and colour losses and large Vitamin C destruction occur with many vegetables. Promising results have been reported for green beans, broccoli, Brussels sprouts, sweet potatoes and pumpkin. In general, radiation-sterilized vegetables do not appear to be superior to those produced by thermal processing. A combination of radiation and heat (8.5 kGy (850 krad) plus five minutes at 100°C), however, has produced a sterile product of quality superior to thermally processed peas. The thermally processed peas were soft and yellow, whereas the irradiated peas were of good texture and green.

As with vegetables, most fruits are damaged by high doses of radiation and are generally unacceptable. Stable apple juice of good quality, on the other hand, can be prepared by a combination of heat and radiation.

Dry spices and related products such as vegetable flakes often contain large numbers of microorganisms which can contaminate the foods to which they are applied. Doses of 10 to 20 kGy (1 to 2 Mrad) are effective in reducing the microbial count of these materials. There is generally good retention of the sensory quality.

Certain dry food materials such as maize starch and rice can also be irradiated effectively to reduce microbial count.

Special diets for hospital patients requiring extraordinary protection from infection are irradiated in the frozen state with doses of about 25 kGy (2.5 Mrad). This procedure is more generally applicable than heat sterilization and provides foods with more acceptable sensory characteristics.

Diets for certain kinds of laboratory animals (specified pathogen-free and germ-free) are also irradiated with doses of about 25 kGy (2.5 Mrad). Irradiation is considered superior to sterilization procedures using heat or ethylene oxide.

7.3. Low-dose irradiation of foods

Low-dose treatments generally result in extension of the shelf-life of the product or destroy contaminating organisms. Very often, irradiation is combined with another preservation method such as refrigeration. Because of the low dose, changes in the sensory characteristics are either too small to be detected or are of minor significance. Flavour changes, in particular, do not generally cause serious difficulties. Softening of some fruit cultivars occurs, however, and precludes the use of irradiation.

7.3.1. Meats and poultry

Raw meats and poultry are of great nutritional value in the human diet and are also of economic importance. Due to the methods used in handling and preparation, meats and poultry become contaminated with bacteria which in time increase in number and spoil the products. In most marketing procedures, such products are kept under refrigeration in order to extend the market life. If the meat or poultry items are treated by ionizing radiation, the bacterial populations will be markedly reduced. Thus the growth of the bacterial populations will be retarded and will not be sufficiently numerous to spoil the product.

Nevertheless, with the best of handling and refrigeration, cold-tolerant organisms will continue to grow and to increase in number sufficiently to eventually spoil the products. It is not economically practicable to use sufficient radiation to kill all the bacteria and prevent spoilage.

The principal spoilage microorganisms of fresh meats and poultry are of the genus Pseudomonas. This group of organisms is relatively sensitive to radiation, having a D_{10} value in the range of 20 to 50 Gy. Hence, relatively small doses of radiation of 500 to 1000 Gy can effectively reduce the population of this meat contaminant to a very low level. The subsequent outgrowth on storage is different from the normal. In the presence of oxygen, the outgrowth is principally gram-negative psychrophilic aerobes, such as Achromobacter and also sometimes certain yeasts. In the absence of oxygen, the outgrowth is primarily due to Lactobacteriaceae.

The use of antibiotics in combination with radiation has been investigated. The Microbacterium species are quite sensitive to certain antibiotics such as the tetracyclines but are relatively resistant to radiation. The antibiotic action supplements that of the radiation and there is a further extension of the life of the product. However, the use of antibiotics in food preservation is not regarded by some regulatory agencies as a desirable practice.

As mentioned in Section 7.2.1, meats and poultry, as well as other foods, develop a foreign flavour upon irradiation. Table XXIII lists the threshold doses which produce a detectable irradiation flavour for a number of protein foods of animal origin when they are irradiated at ordinary refrigerator temperatures. From these values it is clear that there is a fairly low dose limitation for meats and poultry as long as the irradiation is performed at temperatures above freezing.

Fortunately, the doses for radurization of meats and poultry are likely to be below the flavour-threshold values, although this may not be the case for radicidation. For fresh meats and poultry, the principal pathogens of concern are certain species of Salmonellae. A dose of 4.75 kGy (475 krad), yielding a seven-fold decimal reduction, has been proposed. This seems certain to be enough radiation to affect the flavour of the meat unless the irradiation is carried out at sub-freezing temperatures, which may not be possible in all cases.

While irradiation is effective in retarding microbial spoilage of fresh meats at refrigeration temperatures, it does not control colour degradation and the formation of a liquid exudate known as drip or weep. In some circumstances these two degradation processes are of equal significance to microbial spoilage. This is especially true when meats are prepared as retail cuts. In such a case, more than irradiation is therefore needed to preserve fresh meats. One approach to preventing drip and loss of colour is to vacuum-package the meat for at least

TABLE XXIII. THRESHOLD DOSE FOR DETECTABLE OFF-FLAVOUR FOR PROTEIN FOODS FROM VARIOUS ANIMALS IRRADIATED AT 5–10°C

Animal food	Threshold dose[a] (kGy)	Animal food	Threshold dose[a] (kGy)
Turkey	1.50	Turtle	4.50
Pork	1.75	Halibut	5.00
Beef	2.50	Opossum	5.00
Chicken	2.50	Hippopotamus	5.25
Lobster	2.50	Beaver	5.50
Shrimp	2.50	Lamb	6.25
Rabbit	3.50	Venison	6.25
Frog	4.00	Elephant	6.50
Whale	4.50	Horse	6.50
Trout	4.50	Bear	8.75

[a] To convert to krad, multiply by 100 (from J. Food Sci. **37** (1972) 672).

the major part of the distribution period and to treat the meat with condensed phosphates to prevent excessive drip. Phosphates also help to preserve colour. Vacuum-packaging prevents lipid oxidation.

Meats may contain pathogenic parasites (see Table XII). Pork may be contaminated with *Trichinella spiralis*. The reported dose to prevent maturation of ingested larvae of this helminth is between 200 and 500 Gy (20 and 50 krad). Such a dose would not cause detectable sensory changes in pork. Prevention of maturation of the parasite may, however, be an inadequate health measure since the ingested larvae are not killed and might harm the host. Doses to yield an immediate kill of meat parasites are probably greater than 5 kGy (0.5 Mrad). Again, undesirable flavours are likely to result from the dose needed if the irradiation is performed at refrigerator temperatures. Some reduction of dose is possible by holding the irradiated meat for a number of days.

7.3.2. Marine and fresh-water products

The principal reason for irradiating these raw products is to secure the extension of shelf-life by delaying microbial spoilage. It is expected that radiation will be used in conjunction with other preservation agents such as refrigeration.

The value of such extension of life lies in the opportunity it provides for economically transporting these products in a fresh condition greater distances than is now possible.

Marine and fresh-water products are of several kinds and may be classified in more than one way: e.g. fin-fish, molluscs and crustacea; oily and non-oily. As might be expected, the response to radiation will vary with type. Table XXIV lists the maximum dose without detectable sensory change for a number of marine and fresh-water products.

The spoilage organisms are different for the various products. Pseudomonas are the principal contaminants of white-flesh fish and shrimp. Lactobacteriaceae cause oyster spoilage, and Pseudomonas, Achromobacter, Lactobacillus or Corynebacterium are the main spoilage flora of clams.

Figure 21 shows data in the survival of various organisms on Dover sole after irradiation at several levels. Yeasts, micrococci and Achromobacter, although relatively few, were present initially, and survive even 4 kGy (400 krad) in significant numbers. As a consequence of changes in flora, the outgrowth on storage is different from that without irradiation. The method of holding the food also enters into the outgrowth pattern. Under aerobic conditions, Achromobacter can grow and, when present, can be the principal outgrowth organism. Achromobacter spoilage is typical of irradiated fin-fish fillets. When the irradiated product is stored under anaerobic conditions, the principal spoilage organisms are the Lactobacteriaceae. With the changes in flora, the spoilage characteristics are not always typical. Lactobacteriaceae, for example, cause a sour condition quite different from the typical odour of Pseudomonas spoilage.

The best results appear to be obtained with the irradiation of fresh fish. This is due not only to the better microbial condition of the fish when fresh but is also related to hypoxanthine formation in the muscle. This compound is formed in fish muscle on storage and is the product of the disappearance of inosine monophosphate. The presence of this compound is associated with fish of good flavour and is independent of the microbial condition of the product. Stored fish, deficient in inosine monophosphate, is of inferior quality regardless of microbial condition and is therefore less desirable than fresh fish for radiation preservation. This problem is accentuated by the very purpose of the irradiation, namely to obtain extension of shelf-life, and the inosine monophosphate deficiency may affect quality even before the onset of microbial spoilage.

Type-E *Cl. botulinum* is associated with marine and fresh-water foods under certain conditions. It is of special concern because of its ability to grow and produce toxin at temperatures as low as 3.5°C. Holding marine and fresh-water foods for extended periods could represent a hazard from this organism if the temperature were not below 3.5°C. Since commercial handling cannot always guarantee refrigeration as low as this, the actual hazard from higher temperatures has been evaluated by appropriate studies of inoculated product. These studies

TABLE XXIV. MAXIMUM RADIATION DOSE WITHOUT DETECTABLE
SENSORY CHANGE AND SHELF-LIFE EXTENSION UNDER GOOD
REFRIGERATION *(marine and fresh-water products)*

Product	Dose[a] (kGy)	Shelf-life extension (days)
Sea-fish:		
Haddock	2.0−2.5	18
Perch	3.5	18
Atlantic mackerel	3.5	30
Cod	1.5	18
Petrale sole	4.0	25−38
Grey sole	1.0	20
Halibut	4.0	12−23
Pollock	1.5	18
Fresh-water fish:		
Channel cat	1.0	13−18
Yellow perch	2.0	16−20
White fish	3.0	
Molluscs:		
Clams	8.0	
Oysters	2.0	
Crustacea:		
Shrimp	2.0	5−14
King crab	2.0	14−37
Blue crab	2.5	28
Lobster	2.5	10−18

[a] To convert to krad, multiply by 100.

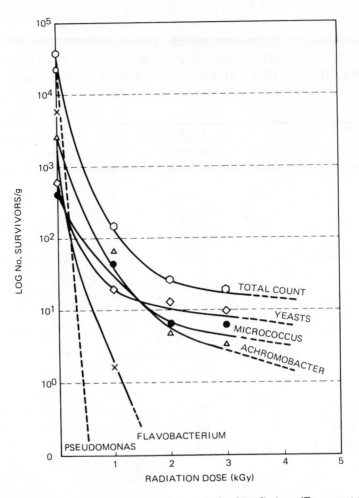

FIG.21. *Microbial flora change in Dover sole as a result of irradiation. (To convert to krad, multiply by 100.)*

suggest that there is variation in toxin formation among the sea- and fresh-water foods, and that each item requires individual study. Holding the temperature below 3.5°C is a uniformly safe practice. Some degree of protection is afforded by the cooking process which ordinarily is sufficient to inactivate any botulinum toxin which may have formed.

Chub mackerel boiled in a saturated salt solution for five minutes will keep without refrigeration for a maximum of three days. Radurization at 2 kGy (200 krad) extends this period to five days. The boiling effect and the 4% salt present are adequate to prevent growth and toxin formation by Type-E *Cl. bot.*

82

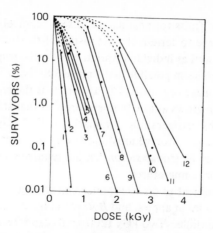

FIG.22. Approximate dose-response curves for spores of post-harvest disease fungi:
(1) Trichoderma viride, *(2)* Phomopsis citri, *(3)* Penicillium italicum, *(4)* Penicillium expansum,
(5) Penicillium digitatum, *(6)* Geotrichum candidum, *(7)* Monilia fructicola, *(8)* Botrytis cinerea,
(9) Diplodia natalensis, *(10)* Rhizopus stolonifer, *(11)* Alternaria citri, *(12)* Cladosporium
herbarum. *(To convert to krad, multiply by 100.) Data from Adv. Food Res.* **15** *(1966) 162.*

Shrimp (*Crangon vulgaris*) cooked in sea-water and irradiated with 1 to 2 kGy
(0.1 to 0.2 Mrad) maintained satisfactory quality under refrigeration for 20 to
24 days, a period about double that without irradiation.

Dried fish may be subject to damage by insects. Doses of 0.15 to 0.50 kGy
(15 to 50 krad) are usually effective in controlling insect infestation of dried and/or
smoked fish. Where the infesting insect is the cheese skipper, *Piophila casei*,
a dose of 2.0 kGy (200 krad) may be necessary.

7.3.3. Fruits

The reason for irradiating fruits may be one or more of the following:

(a) To delay microbial spoilage;
(b) To control an insect infestation;
(c) To inhibit sprouting and delay senescence

Microbial spoilage of fruits is largely concerned with fungi. The desired degree
of fungicidal effect considerably controls the radiation dose employed. For fruits
with a short physiological life, such as strawberries, a temporary halt in lesion
growth may be sufficient and the dose may therefore be relatively small.
With longer-lived fruits such as citrus, a more complete inactivation of fungal
lesions is necessary. Figure 22 shows the approximate doses for inactivation for
spores of the principal post-harvest disease fungi associated with fruits.

The irradiation of fruits for post-harvest disease control is complicated by the possibility of damage to certain characteristics of the fruit, such as texture. Each kind of fruit, as well as individual varieties of fruits, responds differently. In general, useful effects seem possible in the range 500 Gy to 3 kGy (50 to 300 krad).

The principal spoilage organism of strawberries is the grey mould *Botrytis cinerea*. This organism grows at low temperatures, so that strawberry spoilage cannot be controlled by refrigeration. Irradiation with 2 kGy (200 krad) effectively delays spoilage. Storage must, however, be under refrigeration since other organisms, such as *Rhizopus stolonifer,* are relatively radiation-resistant and will grow at higher temperatures.

Some fruits are subject to disease spoilage on storage. *Penicillium expansum* and Gloeosporium species in apples, and *Botrytis cinerea* in pears, are frequently among the fungi responsible. Two kGy (200 krad) can effect a significant reduction of mould damage in apples. Texture changes in irradiated apples on storage are indicated in Fig. 23. It is believed that the immediate softening observed is associated with reduction in molecular weight of pectic substances. Other changes also occur, such as alteration of the organic acid content.

Citrus fruits are subject to a variety of fungi including *Penicillium italicum* (blue mould), *Penicillium digitatum* (green mould), *Phytophthora* spp. (brown rot) and a number of organisms associated with stem-end rot such as *Alternaria citri, Diaporthe citri, Pleospora herbarum, Botryosphaeria ribis* and *Diplodia natalensis.* Up to 2.8 kGy (280 krad) of γ-radiation does not cause damage to the internal quality of Shamouti oranges, but causes pitting of the flavedo or outer layer of the peel. Similar damage has been reported for grapefruit and lemons. Peel damage may be due to radiation-induced diffusion of terpene compounds in exocarp cells. Peel damage can be avoided by reduction of the dose to 1 kGy (100 krad) for oranges and 1.5 kGy (150 krad) for grapefruit when the radiation is combined with heating for about five minutes at 53°C. These doses are effective against the Penicillium blue and green moulds. Alternaria rot, however, may be increased due to radiation-induced death of calyx tissue.

Two kGy (200 krad) control *Monilia fructicola* on peaches but cause unacceptable softening of the fruit. This dose can be reduced to 1 kGy (100 krad) by combining radiation with heating at 50°C for four minutes. Anthracnose infection of mangoes can be reduced by a dose of 1.05 kGy (105 krad) followed by a hot-water treatment. The combination of heat and radiation also appears to be useful for other fruits such as papayas, nectarines and cherries.

The softening caused by radiation can be largely offset by dipping the fruit in a $CaCl_2$ solution. Presumably this restores a calcium-pectin association which is disturbed by the radiation.

Irradiation of ripe tomatoes extends the normal storage period at 22–25°C to as much as six days or longer depending on the level of the initial microbiological infection. Unripe fruit is unsuitable for irradiation (see below).

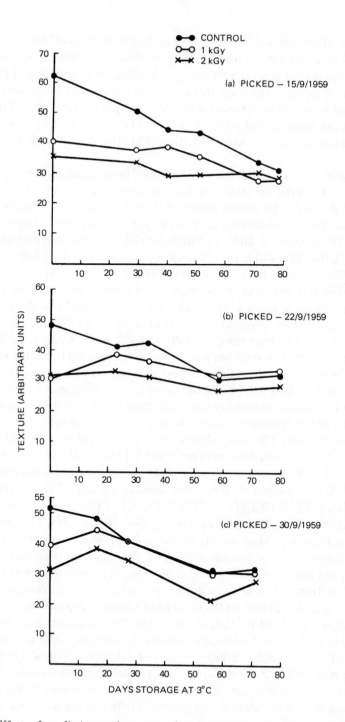

FIG.23. Effects of γ-radiation on the texture of apples. (To convert to krad, multiply by 100.)

Insect infestation is of interest in connection with tropical fruits either fresh or dried. This interest stems from current prohibition of shipment of these fruits into potential market areas as a means of controlling the distribution of the contaminating insects. The fruits that seem to be concerned are citrus (fruit fly), mango (seed weevil, *Sternochetus mangiferae*), and papaya (fruit fly). Fruit fly infestation can be controlled with 0.20 to 0.33 kGy (20 to 33 krad); *Sternochetus mangiferae* can be sterilized with 0.33 kGy (33 krad) and killed with 0.75 kGy (75 krad).

Delay of senescence of fruits may be one of the more important applications of radiation. Of particular interest is delay of ripening of bananas. This fruit is frequently shipped to far distant markets but can be shipped successfully only in the green state. Irradiation in the green state delays the onset of natural ripening. The response of different varieties varies. The Montecristo variety treated with 0.40 kGy (40 krad) will keep in the green state five to six days longer at 26°C. Similar results have been reported with the Gros Michel and Valery varieties. The plantain, or cooking banana, on the other hand, undergoes a greater delay in ripening – as much as nine days. Irradiation of mechanically injured green bananas or of ripe bananas is ineffective in extending their life. Bananas irradiated in the green state ripen normally with ethylene. A second inhibition of senescence at the ripe stage has been reported for bananas irradiated when green. Brown-spotting of irradiated banana skins occurs with some varieties. This is due to increased polyphenoloxidase activity of the skin and pulp. It is attributed to radiation-induced enzyme activation which results from cell-membrane damage and which permits contact between the enzyme and substrate.

A dose of 2 kGy (200 krad) delays the ripening of papayas. A 10% loss of ascorbic acid occurs with doses between 1 and 1.5 kGy (100 and 150 krad). Irradiation of Alphonso mangoes with 0.25 kGy (25 krad) delays ripening for varying periods, depending on the storage temperature. At 20°C the delay for non-irradiated fruit is 10 days; at 5°C it is 40 days. Other mango varieties show similar but not identical delay of ripening. A dose of 1 to 2 kGy (100 to 200 krad) increases polyphenoloxidase activity of mangoes and causes tissue-darkening.

Irradiation of unripe peaches with 3 kGy (300 krad) accelerates carotenoid formation and intensifies anthocyanin formation; as a result it intensifies the colour of the fruit. There is also a slower conversion of pectic substances.

The effect of radiation on the ripening of tomatoes depends on the degree of ripeness of the fruit at the time of irradiation. Irradiation of green fruit leads to reduction of the rate of subsequent carotenoid synthesis. In pink fruit, the synthesis of carotenoids has already begun, and radiation is without effect. In both green and pink fruit, other disturbances of a physiological nature occur, leading to a susceptibility to infection of the tissue with microorganisms during storage. Hence irradiation of tomatoes is best used with fully ripe tomatoes and is therefore limited to control of microbial spoilage.

Fresh Iraqi dates tolerate as much as 5.4 kGy (540 krad) with no objectionable sensory changes. This tolerance has been ascribed to low water activity and low protein content. Irradiation of dates can be used (a) to extend the softening process time, (b) to reduce the time period of after-ripening, (c) to control microbial spoilage, and (d) to disinfest (dried dates). Appropriate doses for these effects are, respectively, (a) 0.1 to 0.3 kGy (10 to 30 krad), (b) 2.7 to 5.4 kGy (270 to 540 krad, (c) 0.9 to 5.4 kGy (90 to 540 krad) and 0.3 kGy (30 krad).

7.3.4. Vegetables

The principal interest in irradiating vegetables has been in delaying senescence through sprout inhibition or similar processes which destroy the acceptability of these raw foods. Of major interest has been the irradiation of potatoes and onions. It has been established that irradiation is an effective means of keeping these foods for extended periods. Other foods, such as mushrooms, show promise of extension of shelf-life but for shorter times.

A dose of 80 Gy (8 krad) is apparently an optimum level for controlling the sprouting of white potatoes, although the dose varies with the variety. Larger doses interfere with suberization (healing of injury) and can lead to increased rotting through microbial invasion of the tissue at locations of injury. The rate of respiration decreases by about 30% immediately after irradiation, followed by a rise to almost the normal level. As shown in Figs 24 and 25, there is an initial rise in reducing sugar and a decrease in ascorbic acid. On storage, such differences from unirradiated potatoes disappear. Gamma radiation prevents internal as well as external sprouting. Some varieties of potato become darker on heating or cooking or on processing such as dehydration. The use of stored irradiated potatoes for processing (e.g. chipping) is satisfactory but (based on a study with the Kennebec variety) problems may be found with extended storage (longer than eight months at 9°C), primarily due to non-enzymatic browning. It is important to store only potatoes of good initial quality, and varietal differences are important. The irradiation storage temperature affects both the quality and quantity of satisfactory potatoes, as shown in Fig. 26. It is usually considered necessary to delay irradiation after harvest until cuts and bruises incurred in harvesting have healed.

Onions respond to radiation somewhat similarly to potatoes. Forty to 80 Gy (4 to 8 krad) appears to be a satisfactory dose range, depending on variety. Discolouration of the interior of the bulb resulting from injury or death of the growing point has been observed. No off-flavours have been detected, but a 'mellowing' of the normal pungency has been reported. Again, varietal differences are important. Onions are best irradiated as soon as possible after harvest.

Inhibition of sprouting by irradiation also occurs with sweet potatoes, yams, garlic, shallots, carrots, red beets, turnips, ginger roots and the tubers of Jerusalem artichokes.

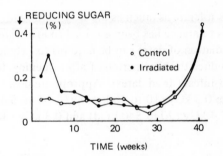

FIG.24. *Effect of storage on mean reducing sugar levels in Kennebec potatoes. After GARDNER, D.S., and McQUEEN, K.F., Atomic Energy of Canada Ltd., The Effect of Gamma Rays on Storage Life and Chipping Qualities of Ontario Crown Kennebec Potatoes, Rep. AECL-2175 and in Potato Chippers (May 1965).*

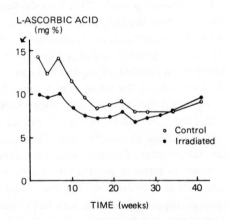

FIG.25. *Effect of storage on mean-L-ascorbic acid levels in Kennebec potatoes. After GARDNER, D.S., and McQUEEN, K.F., as Fig. 24.*

Irradiation of potatoes at 0.1 kGy (10 krad) prevents adult emergence of the tuber moth *Phthorimaea operculella* (Zeller), provided the infestation is with eggs or early larval instars. Late larvae require 0.2 kGy (20 krad).

Mushrooms desiccate and open their caps in storage within five to seven days at 0–4°C. Irradiation of the *Agaricus bisporus* in the dose range 100 to 1000 Gy (10 to 100 krad) delays the cap-opening for 10 to 14 days. The preservation effect of radiation is aided by appropriate packaging to lessen desiccation and gas exchange with the atmosphere.

Radurization of vegetables is concerned primarily with the bacterium *Erwinia carotovora*, although yeasts and moulds may be present. Irradiation increases the life of leafy vegetables such as endive, parsley and cabbage as well as carrots and

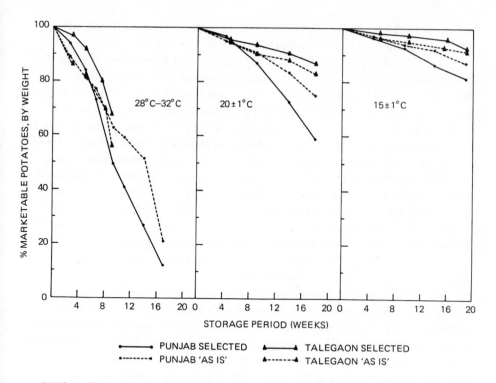

FIG.26. *Marketable percentage of irradiated potatoes as a function of storage time and temperature.*

sliced onions. Irradiation of prepackaged peeled potatoes in combination with a sulphite dip is effective in extending life. Prepackaged soup greens (leek, celery, carrot, cauliflower and onion) irradiated with 1 to 2 kGy (100 to 200 krad) have a life at 10°C of four days, compared with one day without irradiation.

7.3.5. *Cereal grains and their products, leguminous and other beans, animal feeds and baked goods*

The foods in this category generally owe their preservation to their low moisture content. They are, however, subject to various problems for the solution of which irradiation can be helpful. Insect infestations of cereal grains and of leguminous and other beans cause enormous losses. If moisture conditions are appropriate, mould can grow on all these foods. Animal feeds are usually contaminated with bacteria of the Salmonellae group and possibly with other pathogens. Meat, milk and eggs derived from animals consuming such feeds also frequently contain these same microorganisms. Bacteria in some cereal grains and their products affect their life and use. Disinfestation, radicidation and radurization are therefore all applicable to this category of food.

To select the appropriate dose for disinfestation it is necessary to know which insects are involved with the particular product under the conditions of storage or use. While most insects found in stored cereal grains and their products and in leguminous and other beans will be inactivated by 0.5 kGy (50 krad), there can be exceptions. For example, the grain moth *Sitotroga cerealella,* which infests maize and rice, can reproduce after irradiation with doses as high as 0.9 kGy (90 krad). For certain products, less than 0.5 kGy (50 krad) may be sufficient.

It is important to recognize that doses usually given for insect disinfestation are not immediately lethal. In many situations delayed lethality is acceptable; it is of little value, however, in the short-term distribution of foods. Delayed lethality also fails to meet some current government regulatory requirements. It should be recognized that irradiation affords no protection in the event of subsequent infestation and it may therefore be necessary to provide post-irradiation measures such as anoxic storage.

As can be expected, mould control requires relatively high doses. For example, to retard mould growth on wheat with a moisture content of about 15% requires 2.5 kGy (250 krad). A dose of 6 kGy (600 krad) prevents mould growth on maize and milo. Mould growth on cocoa beans stored at 95% relative humidity and 26°C is inhibited by 5 kGy (500 krad). This dose can be reduced to 0.5 kGy (50 krad) by pretreating the beans with 100°C air for 15 minutes.

Packaged sliced bread irradiated with 5 kGy (500 krad) remains free of mould for at least 11 weeks. Here too, heat may be used to reduce the dose requirement. Irradiation at 65°C with a dose of 0.5 kGy (50 krad) provides an equivalent result. Indian unleavened bread (chapaties) remains free of mould for more than six months when irradiated with 10 kGy (1 Mrad).

Production of aflatoxin by *Aspergillus parasiticus* on bread is prevented by a dose of 2 kGy (200 krad).

Some cereals and their products, such as rice and maize starch, contain bacteria whose presence affects product quality and use. Doses of about 3 kGy (300 krad) reduce the microbial count adequately.

The radicidation of animal feeds for the purpose of eliminating human pathogens, such as Salmonellae bacteria, requires a fairly large dose. A dose of 7.5 kGy (750 krad), designed to effect a five-decimal reduction of Enterobacteriaceae in animal feeds, has been proposed.

The irradiation of wheat causes fragmentation of the carbohydrate and protein components to lower molecular weight entities. These changes affect the baking qualities of wheat flour by facilitating amylolytic and proteolytic action. For doses up to 2 kGy (200 krad) the loaf volume and texture of leavened bread is improved. Doses greater than 2 kGy (200 krad) lead to lowered bread quality. Unleavened chapaties, on the other hand, are of satisfactory quality

TABLE XXV. THE EFFECT OF EGG PRODUCTS ON THE SENSITIVITY OF STRAINS OF SALMONELLAE TO IRRADIATION WITH HIGH-VOLTAGE CATHODE RAYS

Product	Serotype	D_{10} (kGy)[a]					
		Liquid		Frozen		Dried	
Whole egg	S. typhimurium	0.40		—		0.557	0.498
	S. senftenberg	0.17		—		0.605	0.450
Yolk	S. typhimurium		—	0.427	0.534	0.759	0.665
	S. senftenberg		—	0.379	0.486	0.806	1.03
White	S. typhimurium	0.338	0.403	0.356	0.249	0.806	0.865
	S. senftenberg	0.243	0.308	0.160	0.190		
White, sugared	Both					1.10	1.30

[a] To convert to krad, multiply by 100.

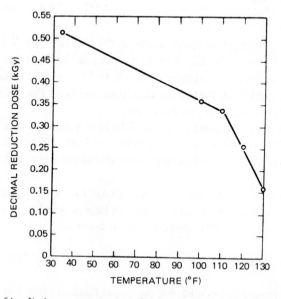

FIG.27.   Effect of irradiation temperature on the decimal reduction dose of S. typhimurium in whole egg. Data derived from J. Food Sci. **29** (1964) 469. (Original measuring points were in degF and are not converted to degC).

with doses as great as 10 kGy (1 Mrad). Flavour differences have been observed in bread and cakes made from wheat irradiated with 0.2 kGy (20 krad). A dose of 2 kGy causes a darkening of the crust colour, probably the result of a Maillard reaction.

The dose limit for acceptability of brown rice is 1 kGy (100 krad) and for milled rice 2.5 kGy (250 krad).

Dry 'navy' beans irradiated with 1 kGy (100 krad) have acceptable sensory characteristics. Owing to increased tenderness, cooking time is reduced. A similar softening occurs with red gram.

Coffee beans treated with a dose of 1 kGy (100 krad) for insect disinfestation yield brewed coffee of acceptable flavour and odour.

7.3.6. Eggs

The principal interest in the irradiation of eggs has been to remove the health hazard related to their Salmonellae content. The availability of heat-treatment methods to solve this problem with eggs has substantially lessened the interest.

It has been observed that there are differences in the radiation-resistance of strains of Salmonellae and that the resistance of a given strain varies with the nature of the egg product in which it resides. Table XXV shows data illustrating this variation for several serotypes of Salmonellae in different egg products.

It has been suggested that for Salmonellae an inactivation factor of 10^5 to 10^7 is required in eggs. Using the highest D_{10} value 0.403 kGy (40.3 krad) (for S. typhimurium (Table XXV)), a dose of about 3 kGy (300 krad) is indicated for liquid egg-white.

The effect of irradiation temperature on the D_{10} value for S. typhimurium in whole egg is shown in Fig. 27. A sharp decrease in the D_{10} value occurs at about 45°C. It is probable that the temperatures above 45°C exert a significant bactericidal effect, apart from that of the radiation, and form the basis for the 'combined treatment' of heat and radiation.

Cakes baked from irradiated eggs tend to have a slightly reduced volume. The stability of the foam prepared from irradiated fresh and frozen egg-white is somewhat less than that of comparable unirradiated products, but depends on the dose used.

Irradiation of shell-eggs causes severe damage to the thick white, giving the egg an 'old' appearance. Yolk membranes can be weakened or broken. Irradiation of shell-eggs has therefore not been considered feasible.

7.4. Miscellaneous applications

Frozen horse-meat, intended for use as pet food, is often found to contain Salmonellae. Based on D_{10} value of 1.28 kGy (128 krad) for the most radiation-resistant strain, S. typhimurium, a dose of 6.4 kGy (640 krad) has been proposed for this material.

TABLE XXVI. DOSES FOR SELECTED DEHYDRATED VEGETABLES AND FRUITS TO REDUCE COOKING TIME ON REHYDRATION

Product	Approximate preferred dose (kGy)[a]
White onion flakes	3
Tomato flakes	6
Potato dice	10
Carrot dice	20
Dried fresh peas	20
Leek	24
Bell pepper dice (green or red)	25
Cabbage flakes	30
Green lima beans	30
Celery flakes	33
Cut green beans	40
Okra pieces	40
Beet cubes	>40
Apples	50
Prunes	90

[a] To convert to Mrad, divide by 10.

"Nham", a Thai fermented pork product, eaten raw, can be freed of Salmonellae by irradiation with 2 kGy (200 krad).

Fresh French prunes irradiated with 4 kGy (400 krad) or more, and then dried, were found to be tenderer than similar unirradiated dry prunes. Such irradiated prunes generally took up more water on reconstitution. This same amount of radiation shortened the freeze-drying time of prunes by about one third. Blueberries irradiated with 3 kGy (300 krad) lost 83% of their original weight in 30 hours as compared with 72% for unirradiated berries. Navy beans treated with 4 kGy (400 krad) had the highest water uptake under a variety of rehydration conditions.

The softening effect of radiation has been considered as a means of reducing the cooking time for dehydrated vegetables in soups. Since different vegetables react differently to radiation, different doses are needed. Moreover, different parts of a given vegetable react differently. Skins of lentils or lima beans, for example, are preferably softened and a more uniform texture of these foods is obtained through irradiation. With tomatoes, however, the skin is more affected than the flesh, causing an accentuation of the textural differences of these parts. The gum of okra, which accounts for a major characteristic of this vegetable, is destroyed by radiation.

Table XXVI lists the doses required for selected dehydrated fruits and vegetables in order to reduce the cooking time during rehydration from between ten and twenty minutes to one or two minutes. Any deleterious effects on flavour or appearance are considered minor.

Yield in juicing operations can be increased by irradiation. With grapes, the yield of juice is increased 2−28% in proportion to dose in the range of 0.5 to 16 kGy (0.05 to 1.6 Mrad). The yield of date syrup (dibis) from dates in the fully soft (rutab) stage is also similarly increased.

The oligosaccharide content of soybeans can be reduced by irradiation and controlled germination. A dose of 2.5 kGy (250 krad) is applied 24 hours after germination. The beans are incubated an additional 72 hours and are then air-dried. Raffinose and stachyose contents are reduced by approximately 75%.

Malting losses may be reduced 1−2% by irradiating dry barley with doses of 0.5 to 8 kGy (50 to 800 krad).

8. PACKAGING

The reasons for packaging food can be manifold, but the basic purpose is to protect it from the environment.   If the radiation treatment is intended to control microbial spoilage, then a very large area of this environmental protection is the prevention of recontamination of the food. In other cases the technical function of the package may be to prevent moisture loss or moisture uptake, to provide an atmosphere other than air, to protect the food from mechanical damage, or simply to keep it clean.

The packaging of irradiated foods is somewhat unusual in that in most cases the food can be packaged before treatment. Only low-energy electrons or certain X-rays would present problems of penetration. Under proper conditions, irradiation in the shipping or bulk container is possible.

The effects of radiation on the principal materials used in packaging are shown in Fig. 28. This information suggests that irradiation can be used successfully with most conventional packaging materials. The following additional information on the classes of packaging materials maybe a guide to their selection:

(a)   Cellulose. This is a natural polymer with a high molecular weight and a crystalline character. In addition to natural cellulose, a variety of cellulose derivatives are available, such as cellophane, rayon and cellulose acetate. Irradiation

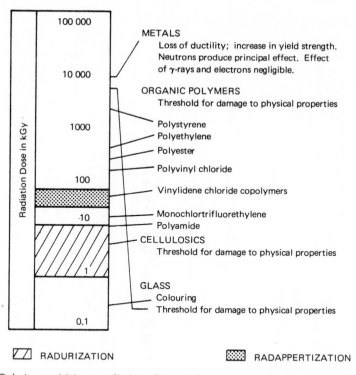

FIG.28. *Relative sensitivity to radiation of principal materials used in packaging. There is a wide variation in the effect of irradiation on various materials, and radiation levels are approximate. (To convert to krad, multiply by 100.)*

causes chain degradation and other chemical changes, leading to decrease in molecular strength. The cellulose polymers are among the most radiation-sensitive packaging materials.

(b) Glass. In glass, irradiation produces free electrons, which may be trapped and cause the formation of colour-centres. At high doses this causes the (uncoloured) glass to turn brown. Heating the glass will restore the initial colourless condition. Radiation produces no other significant effect on glass.

(c) Metals. Radiation at the energy level employed in food irradiation increases the mobility of the outer-shell electrons of metals. This added energy ultimately degrades to heat, of negligible quantity. There is no other effect on metals.

(d) Organic polymers. Free radicals are formed in polymeric substances and these lead to cross-linking or to chain scission, or to recombination. Hydrogen and other chemicals may be formed at the same time. If oxygen is present, oxidation may occur. The final result is determined by the predominant reaction, which is dependent on the chemical structure of the polymer, the stress and environment.

TABLE XXVII. HEADSPACE GAS IN CANNED FOODS IRRADIATED AT 45 kGy (4.5 Mrad)
(can size: 307 × 409; storage: 5–7 years)

Product	Enzyme activity	Total gas (mltr)	N_2	O_2	H_2	CO_2	CO	CH_4	H_2S
						%			
Chili	Inactive	39.6	4.6	0	85.5	5.3	1.7	2.6	0
Cherries	Inactive	22.0	3.9	0	86	8.2	1.8	0	0
Green beans	Inactive	56.0	17.0	0	75.8	7.0	0	0.2	0
Ground beef	Active	80.0	32.9	1.3	29.3	31.0	2.2	3.2	0
Ground beef	Inactive	25.0	30.5	0.4	48.3	17.8	1.4	1.6	0

From J. Food Sci. **32** (1967) 201.

TABLE XXVIII. HEADSPACE GAS IN WHOLE BONED HAM IRRADIATED AT 45 kGy (4.5 Mrad)

Container	Storage time (months)	Total gas (mltr)	N_2	O_2	H_2	CO_2	CO	CH_4	H_2S
						%			
Polyethylene bag	4	80	87.7	9.0	0	0	1.6	1.7	0
Polyethylene/foil laminate bag	4	40	81.0	2.0	0.01	0	11.6	5.4	0
No. 10 can	79	800	35.7	0	46.5	17.8	0	0.02	0

From J. Food Sci. **32** (1967) 201.

Radiation-induced changes can have a great effect on the physical properties of the polymer. Cross-linking can increase tensile and flexural strength and decrease elongation, crystallinity and solubility. Shortening of the polymeric chain through scission results in a decrease in tensile and flexural strength.

The choice of the packaging material and the nature of the container for a specific food is usually determined by the purpose they are to serve. Sterilized food must have a container which prevents access to bacteria and other micro-organisms. This means a tightly closed container. Packages for low-dose applications do not need to be tight.

Irradiation causes gases to form which can cause tight containers to swell. Tables XXVII and XXVIII show the results of analyses of gases in the headspace of various irradiated foods. Table XXIX gives similar data for model systems. The effect of radiation temperature on headspace composition is shown in Table XXX. Since most flexible polymeric films are permeable to hydrogen, this gas disappears from packages made of such films on storage.

8.1. Rigid containers

The only rigid primary container for irradiated foods studied so far has been the metal can. Steel containers, tin-plated and lined with an appropriate enamel, have proved satisfactory. Table XXXI indicates the enamels found to be satisfactory for several foods.

Sulphur compounds from irradiated foods do not react with the metal surface of the can as readily as those from thermally processed foods. Irradiated foods, however, have marked dezincing properties, and zinc oxide pigments should not be used in enamels.

Three end-sealing compounds have performed satisfactorily. They are: (1) blends of cured and uncured butyl elastomers; (2) neoprene and butadiene-styrene elastomers; and (3) neoprene and uncured butyl rubber elastomers.

Aluminium cans are of interest because of their relatively low density and consequent smaller absorption of radiation. Such cans perform satisfactorily with appropriate enamels. The radiation-generated gas, however, can be a problem owing to the relatively smaller physical strength of the aluminium can compared with that of the steel-based can. Owing to the generation of gas, some underfill of either type of can is desirable.

Secondary rigid containers such as shipping or bulk containers made of fibre-board (base material cellulose) suffer some loss of protective characteristics, but are generally satisfactory.

8.2. Flexible containers

Flexible plastic containers offer the means of saving weight and cubic space. In addition, their low density makes them attractive for use with irradiated foods.

TABLE XXIX. HEADSPACE GAS IN MODEL SYSTEMS REPRESENTING INDIVIDUAL FOOD COMPONENTS IRRADIATED AT 45 Gy (4.5 Mrad) (can size: 303 × 406)

Model system	Total gas (mltr)	%					
		N_2	O_2	H_2	CO_2	CO	CH_4
Water (distilled)	5.0	37.0	1.1	58.7	3.3	0	0
Water + 2% NaCl	5.5	29.4	0	68.6	2.0	0	0
Water + 10% sucrose	50.0	9.2	1.6	82.5	4.1	2.5	0
Water + 10% starch	50.0	16.2	2.5	78.5	1.1	1.8	0
Water + 10% dextrose	58.0	9.2	1.3	81.2	6.6	1.7	0
Water + 6% gelatin	40.0	8.4	0	65.1	0.3	22.8	3.4
Water + 10% corn oil	15.0	17.5	0	79.4	0.4	2.4	0.3
Sucrose (dry)	162	13.4	1.8	84.0	0	0	0
Starch (dry)	170	8.8	0	71.6	4.8	14.9	0
Dextrose (dry)	170	12.9	1.1	83.7	2.3	0	0
Gelatin (dry)	45	86.2	6.8	1.2	0.1	2.5	3.3
Oil (dry)	60	61.4	1.8	33.5	0	9.1	0.1

From J. Food Sci. **32** (1967) 202.

TABLE XXX. EFFECT OF IRRADIATION TEMPERATURE ON GAS PRODUCTION IN SUCROSE AND GELATIN SOLUTIONS IRRADIATED AT 45 kGy (4.5 Mrad) *(can size: 303 × 406)*

Irradiation temp. (°C)	Total headspace gas (mltr)	
	10% sucrose solution	6% gelatin solution
20	58	35
5	62	32
-40	31	23
-80	29	23
-196	23	21
Control (unirrad.)	5	4

From J. Food Sci. **32** (1967) 203.

TABLE XXXI. CAN-ENAMELS SATISFACTORY AFTER 12 MONTHS STORAGE AT 37°C

Product	Enamel	
	Best	Alternate
Cherries	Polybutadiene	Oleoresinous
Chili	Polybutadiene	Epoxy-phenolic
Beef	Polybutadiene	Epoxy-phenolic
Codfish	Epoxy-phenolic	Heat-reactive phenolic
Pork	Polybutadiene	Epoxy-phenolic

Films thinner than 25 μm have too many imperfections to be suitable for use. Films 25 to 76 μm thick are proof against microorganisms, but creasing will damage such films sufficiently to make them unsatisfactory. Films over 76 μm thick appear satisfactory.

The most satisfactory flexible package developed so far for radiation-sterilized food employs a three-ply laminate made up as follows: 25 μm Nylon 6 (outside), 9 μm aluminium foil (middle) and 62 μm intermolecularly bonded polyethylene terephthalate medium-density polyethylene. The polyethylene is the food-contacting layer. This lamination appears to provide the best performance based on the following criteria: that the films are not changed adversely:(a) in their protective characteristics (e.g. heat stability, permeability, etc.); (b) by radiation-induced changes in the food; (c) causing transmission of toxic or potentially toxic substances to the food.

Flexible packages for radiation-pasteurized haddock fillets were found to be satisfactory when made from the following films:

> Saran-coated Nylon 11
> Nylon 11
> Polyolefin-coated polyester
> Semi-rigid polystyrene
> Paper-aluminium-polyolefin-coated polyester
> Nylon-Saran-coated polyethylene
> Aluminium-coated Nylon 11
> Aluminium-paper-polyolefin-coated polyester
> Polyethylene-coated Nylon
> Nylon-Saran-polyethylene

Films of polyethylene and polypropylene were not satisfactory.

For radiation-pasteurized fresh meats, the following oxygen-permeable films appear satisfactory:

> Polyvinyl chloride (fresh meat wrap)
> Cellophane (fresh meat type)
> Polyethylene (high oxygen permeability type)

Among oxygen-impermeable films satisfactory for use with meats are:

> Polyvinylidine chloride (Saran)
> A laminate of polyvinylidine chloride, polyester and polyethylene

Flexible packages for other irradiated foods have not yet been developed in any special way. It is likely that existing information as indicated above can be used as a guide.

9. COMBINATION PROCESSES

It may be desirable to reduce the dose requirement either because the dose needed to attain a particular objective may in some cases be more than the food can tolerate, or for other reasons. One way to do this is to use radiation in combination with some other agent, such as heat or refrigeration. The preceding sections contain a number of such instances.

Heat and radiation may be used in separate sequential steps. In treating foods to kill bacteria, the combination of heat and radiation generally reduces the requirement of each when used separately. There is evidence, however, that synergism occurs only with the sequence, *radiation followed by heat,* and not by the reverse order.

In some situations, heat and radiation may each be used for different purposes. In treating certain fruits, for example, irradiation may be used for insect disinfestation, while heating may be employed to kill spoilage fungi. Without the use of heat in such cases, the dose for fungi kill as well as for insect disinfestation may be more than the fruit can tolerate.

Irradiation at high temperatures leads to reduced dose requirements. This is demonstrated by the reduction in D_{10} value for the spores of *Cl. bot.* with rising temperature, as shown in Fig. 14. A similar effect is shown by *Salmonella typhimurium.* This combination process has been called thermoradiation.

Radiation can be used with other agents, in addition to heat. Shrimps have been dried to 40% moisture and irradiated with 2.5 kGy (250 krad). Compression increases the lethality of radiation. Chemical additives can be used in combination with radiation. Radurized foods are also frequently refrigerated.

10. LIMITATIONS OF FOOD IRRADIATION

The advantages and accomplishments of food irradiation have been covered in the previous sections. Food irradiation, however, is not a panacea for all food problems. While there is little value in listing all the things irradiation cannot do, it is worth noting some of the important limitations which result from technical effects. (Economic factors which limit food irradiation in general will not be considered.)

Not all effects of radiation on foods are desirable. Sensory changes, for example, may severely limit irradiation. Of the sensory changes, the production of 'irradiated flavour', or odour, is probably the most significant. This frequently constitutes a dose limitation for which there may be no satisfactory method of circumvention. The only effective method available at present for reducing the irradiated flavour intensity is to irradiate at sub-freezing temperatures. For either technical or economic reasons, however, some foods cannot be frozen.

The most important limitations for fruits and vegetables are probably softening and related texture changes. It is frequently impossible to use even moderate doses, such as are needed to kill spoilage fungi, owing to the radiation-induced depolymerization of carbohydrate substances that are the basis of the normal texture of these foods.

Colour changes tend, on the whole, to be of minor significance. Colour changes associated with the activation of an enzyme system, such as occurs in some fruits, are probably the most important. Impairment of functionality can be a serious undesired change. In a food such as shell-egg, it largely precludes the use of irradiation; in wheat flour, it clearly limits the dose. Nutritional losses may be regarded as a limitation on food irradiation, but since they are in general small, they are not a significant limitation. Differences in response to radiation related to varietal differences of fruit and vegetables are sometimes important, especially where radiation-sensitive varieties have a good market acceptance.

Some biological effects can restrict the use of irradiation. When a living food is irradiated, it is impossible to direct the radiation to only a single biological process. For example, in inhibiting the sprouting of potatoes, radiation also inactivates the suberization process. This limits irradiation for sprouting inhibition only to intact potatoes, free of wounds. Processes whose normal completion may be desired may also be affected. In treating green tomatoes for control of spoilage microorganisms, the ability of the tomato to develop the normal red colour is destroyed. In such situations, irradiation is not a selective process.

Certain results cannot be secured through irradiation, and in these circumstances it is inapplicable. For example, at doses within the practical limits, irradiation does not inactivate viruses, enzymes, bacterial toxins and mycotoxins. Similarly, it is not immediately lethal in insects. Irradiation has no persistent effect. Post-irradiation contamination can therefore cause spoilage or other damage and must be prevented in order to secure the benefit of irradiation. In some circumstances it is impossible to prevent re-infestation. For the latter, the repeated use of radiation is regarded as acceptable to low-moisture-content food commodities provided that no significant impairment of nutritional or technological properties occur, and where insect re-infestation could not be effectively prevented under practical conditions of storage and transport.

Irradiation is generally directed towards action on living organisms. It does not provide protection for a food by controlling chemical or physical deterioration.

11. WHOLESOMENESS OF IRRADIATED FOODS

11.1. General

'Wholesomeness' is a term used to indicate that a food has acceptable nutritional quality and is toxicologically and microbiologically safe for human consumption. It must be recognized that no food is completely safe in all circumstances; some risk, however small, always exists. The requirement is that this risk be infinitesimal. That risk is balanced by the benefit gained. The judgement of the risk-benefit situation determines whether the food is acceptable or not.

Proof of safety, up to a certain point, is proof of the absence of a hazard. Here too, it is impossible to obtain 100% assurance of the absence of risk. Procedures for determining the risk are specific for the kind of risks involved. Such procedures are being developed continuously and new information can alter an earlier judgement on safety.

It is generally accepted that no known hazard to health should be introduced in the use of irradiated foods. As with other methods of food preservation, irradiation can lead to certain biological, chemical and physical changes in the treated food. Therefore, in order to determine whether such possible changes can create a health hazard to the consumer, it is important to evaluate irradiated foods in terms of their wholesomeness.

In the early work on evaluating the wholesomeness of irradiated foods it was considered necessary to examine each and every irradiated food by separate studies. While specific effects may be related to particular foods, there is evidence from radiation chemistry that foods of similar composition have a generalized, largely non-specific response to radiation. Because of this, individual testing of foods is now considered to be less necessary.

An important point in the findings of radiation chemistry is that the amounts of radiolytic products, even at radappertization doses, are small. The data of Table XXXII show that the amounts of classes of radiolytic volatile compounds found in radappertized beef total approximately 30 ppm. For a given dose and irradiation temperature, the actual amount formed varies with the composition of the beef, particularly the fat content. About 65 individual volatile compounds have been identified in radappertized beef, and it is clear that the amount of any one compound cannot be large.

That the amount of radiolytic products must be small can be seen from consideration of the amount of energy employed in food irradiation. If one takes 50 kGy (5 Mrad) as the dose for radappertization (the probable largest dose to be used), it can be recognized as a small amount of energy, roughly 11 calories per gram. If this ionizing energy were completely effective in breaking chemical bonds, which it is not, it could break only 0.01% of the bonds present in a typical food. Actually, most of the absorbed energy degrades to heat. Therefore, the chemical change in a food caused by irradiation must be small.

Data such as are shown in Fig. 29 demonstrate that the amounts of radiolytic products increase with dose. At different doses, the same products are formed and only the amounts are different. It is therefore possible to extrapolate from one dose to another when evaluating the wholesomeness of irradiated foods. The administering of foods irradiated at doses greater than will be used in practice has

TABLE XXXII. ABUNDANCE OF VOLATILE COMPOUNDS IN RADAPPERTIZED BEEF

Compound	ppm
Alkanes	12
Alkenes	14
Alkanals	1.5
Sulphur compounds	1.0
Alkanones	<0.5
Alkylbenzenes	<0.1
Esters	<0.1
Approximate total	30

From J. Agric. Food Chem. 23 (1975) 1038.

been objected to because excessive radiation can alter the sensory acceptance, possibly makes the food unsuitable for human consumption, and may introduce difficulties in the conduct of animal-feeding experiments.

With something as complicated as a food, it is not possible to identify for certain all the substances that result from irradiation. Were it possible, it might not be necessary to test the food itself; each identified substance could be evaluated by standard toxicological procedures. Because of this difficulty, the method used has been to give the irradiated food to test-animals and to observe their biological reactions and behaviour.

In using standard toxicological and pharmacological procedures in evaluating the wholesomeness of irradiated foods one difficulty is that it is impossible to feed test-animals at a dietary level that exaggerates the normal intake to the extent necessary for evaluating an isolated defined substance. Only the maximum amount of a food can be administered that can be tolerated by the test-animal and does not cause nutrition problems or alter the biological responses.

Multigeneration and long-term feeding studies require large numbers of animals; they usually take years to complete, generate large ancillary efforts such as pathology examinations, are expensive, and do not detect subtle hazards due to weakly toxic substances present in low concentrations. Better and shorter evaluation procedures have therefore been sought. A number of such methods are available, particularly for predicting carcinogenicity and mutagenicity.

Concern has been expressed about the validity and relevance of the interpretation of the results of carcinogenicity and mutagenicity tests. As with all

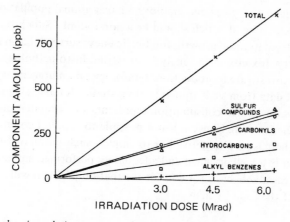

FIG.29. *Graph showing relative amounts of component produced in beef irradiated as a function of dose at $-185°C$ (ppb = parts per billion, where 1 billion $= 10^8$). (To convert Mrad to kGy, multiply by 10.) Reprinted from J. Agric. Food Sci. 23 (1975) 1039.*

animal-feeding tests, they involve extrapolation from laboratory-animal species to man. A greater concern about this extrapolation arises from the use of in-vitro procedures or non-mammalian species in some tests. The absence of defence mechanisms comparable with those of mammals, including man, is an important factor. Despite these difficulties, the accumulation of results of various kinds of testing is considered to be helpful in evaluating wholesomeness.

For irradiated foods the following areas are considered pertinent for evaluation of wholesomeness:

General toxicology	Nutrition
Carcinogenicity	Microbiology
Teratology	Induced radioactivity
Mutagenicity	Packaging

Some of these areas can be evaluated by chemical or physical methods. Information in other areas can be secured only by animal studies or appropriate microbiological investigations.

11.2. General toxicology

For the general toxicological area, subchronic and chronic animal-feeding studies have been employed. Protocols for subchronic tests have required the supply of control and irradiated foods to animals at an acceptably high dietary level for at least 10% of their life span. The subchronic test is considered an appropriate challenge to detect an abnormal biological response. The chronic study usually lasts at least two years and employs a larger animal population, usually in at least two species, one of which should be a non-rodent. Subchronic and chronic studies yield information on growth, food efficiency, survival, haematology, clinical chemistry, toxicity, urine analysis, gross and histo-pathology. Other studies are designed to investigate reproduction, teratology and mutagenicity.

The use of data from such animal-feeding studies for assessing the safety of irradiated foods for human consumption represents an extrapolation from laboratory species to man. Since it is not possible to conduct comparable studies with humans, there is no alternative but to employ such extrapolation. The absence of a toxic effect in a well designed and adequately performed animal-feeding study can provide a reasonable basis for confidence that no problem will arise through consumption of the food by man.

11.3.   Carcinogenicity

Carcinogenicity studies are likewise made in laboratory animals. It is customary to use two laboratory rodent species of known biological responsiveness. Rats and mice are convenient species for these studies.

11.4.   Teratology

Teratological investigation involves the determination of the number and kinds of foetal abnormalities induced by giving pregnant females irradiated compared to unirradiated food.

11.5.   Mutagenicity

A number of methods have been suggested for investigating the mutagenic potential of irradiated food. No single generally accepted method exists which will give an unequivocal result. Progress in this field has been rapid and the selection of tests has hitherto been rather arbitrary and based on knowledge existing at the time.

The *dominant lethal test* has been used as an in-vivo test to reveal chromosomal alterations due to mutagens suspected to be present in irradiated food. Male animals are given the irradiated food for a specified period and mated with untreated females. The pregnant females are examined for pre- and post-implantation losses. The test gives no clue to the mutagenic mechanisms involved and is regarded as rather insensitive for detecting weakly mutagenic substances. It has the advantage of using the intact animal.

The *host-mediated assay* has also been used as an in-vivo test for detecting mutagenic metabolites. It involves the use of an indicator organism, such as particular strains of *Salmonella typhimurium* or other organisms, which are injected into the peritoneal cavity of the mammalian host, such as the mouse. After a time,

106

the indicator organisms are withdrawn and examined for the presence of mutants. The host-mediated assay has now been largely abandoned as a satisfactory test for detecting mutagenicity because of its failure to detect many powerful mutagens and its poor correlation with other more sensitive procedures.

In-vivo *cytogenicity tests* have been employed to detect chromosomal damage. The test-animal is given irradiated food for several days. Cytogenetic analyses of cells from various important target sites are then examined for evidence of chromosomal abnormalities. Most commonly, bone-marrow, peripheral lymphocytes and germ cells are studied. This test is reasonably sensitive and is regarded as an acceptable in-vivo procedure.

In additon to these in-vivo methods, in-vitro tests may be used, of which the induction of back mutations (Ames test) in certain bacterial strains is the most frequently used procedure. This test detects back mutations induced in histidine-requiring mutants of *Salmonella typhimurium.* For individual defined chemical substances, a positive Ames test has a high correlation with carcinogenicity. The test is difficult to apply to irradiated food per se, but modifications have been developed to overcome this problem.

11.6. Nutrition

The nutritional quality of irradiated foods can be assessed by taking into account the following:

Vitamin content, stability and physiological availability;
Fat content, quality and essential fatty acid composition;
Protein quality;
Digestibility of fat, carbohydrate and protein components of the food,
    and the availability of the potential biological energy derived from them;
The absence of antimetabolites;
The subjective qualities of food that make it desirable to eat.

Some of these factors can be evaluated through chemical analyses. The best way to evaluate all nutritional components collectively is through an animal-feeding study measuring items such as growth, reproduction, food consumption and efficiency, and the occurrence of gross abnormalities.

11.7. Microbiology

The area of concern for the microbiological aspects of irradiated foods will vary with the kind of treatment involved. For sterilized products, the dose must be adequate to destroy or inactivate all spoilage microorganisms. For low-acid foods with high water content and which allow germination of the spores of *Cl. botulinum,* the dose must be sufficient to accomplish a reduction of the spore count by 10^{12}, i.e. by 12 D_{10}. As has been noted, this calls for about 45 kGy (4.5 Mrad) for Type-A *Cl. botulinum,* the most radiation-resistant type. To be absolutely safe on this important point, an 'inoculated pack study' is usually employed. In this technique the food in question is inoculated with appropriate levels of the spores of *Cl. botulinum,* processed (including irradiation at several levels) and stored under conditions which will allow spore germination. Growth and toxin formation are correlated with radiation dose, and the minimum dose requirement is established.

For products irradiated with a less than sterilizing dose and for which a microbial spoilage eventually occurs, there are other important microbiological considerations. The radiation may eliminate or inhibit the normal outgrowth of the usual flora, leading to a different microbial spoilage pattern. This new pattern must be identified and evaluated to determine whether it causes a possible health hazard to the consumer of the food. The new outgrowth could include substantial numbers of a particular organism, which is usually suppressed by the flora normal to the unirradiated food.

11.8. Induced radioactivity

The view is generally accepted that any added radioactivity is undesirable and is not to be permitted. For this reason, as discussed in Section 1.4.5, the only radiation sources permitted are ^{60}Co, ^{137}Cs, up to 10 MeV accelerated electrons, and X-rays from a source producing a beam of energy not higher than 5 MeV. All these sources have energy levels below that necessary to induce radioactivity in elements contained in foods.

11.9. Packaging

The basic consideration of packaging in terms of wholesomeness is the transfer from the package to the food of any material capable of causing a health hazard to the consumer. A secondary consideration is the ability of the package to protect the food from the environment. These same considerations apply to any package used for food. The only new requirement is the effect of radiation on the food or on the material from which the package is made.

11.10. Results of testing for wholesomeness

A great deal of effort has been expended on evaluating the wholesomeness of irradiated foods, and work is continuing. As previously mentioned, the early work largely involved testing individual foods by animal-feeding studies. More recent work has been directed to (1) greater use of the knowledge of the radiation chemistry of foods, (2) evaluation of irradiated complete diets, and (3) treatment of irradiation as a general food process instead of considering it as a food additive.

The sum of valid completed investigations of wholesomeness has not demonstrated positive evidence of a toxicological or microbiological hazard associated with the consumption of any irradiated food. The normal nutritive values of the macronutrients of foods (protein, lipid and carbohydrate) are maintained. In some uses of irradiation, certain vitamins are labile to radiation, but the losses are either small or comparable to those associated with other commonly used food processes. A number of food-contacting packaging materials of all needed types have been found to cause no health hazard when irradiated.

The problem of induced radioactivity is resolvable on the basis of proper control of the energy level of the radiation applied to food. This solution of the potential problem of induced radioactivity has been arrived at through studies which included: (a) the irradiation of elements in foods expected to produce radioactive products through isomer activation requiring low energy (γ, γ' reaction) and the particle emission reaction requiring higher energy; (b) the irradiation of foods enriched with these elements; and (c) the irradiation of unenriched foods. Sources employed were ^{137}Cs, ^{60}Co, spent fuel rods, and electrons and X-rays with energies of 4 to 25 MeV. From these investigations it was concluded that:

(a) The commonly used isotopic sources (^{60}Co and ^{137}Cs) emitting radiation of a maximum energy of $\leqslant 1.33$ MeV and X-rays up to 5 MeV do not induce radioactivity;

(b) No detectable induced radioactivity is found in foods irradiated with electron beams of less than 10 MeV, and the induced activity is negligible and very short-lived below an energy level even as high as 16 MeV.

Present views are that γ-rays and X-rays up to 5 MeV and electrons with energies up to 10 MeV can safely be used for food irradiation.

12. GOVERNMENT REGULATION OF IRRADIATED FOODS

From the point of view of consumer safety and in order to protect consumers against false representation of what they are purchasing, it is generally agreed that governments should regulate the manufacture, storage and distribution of irradiated foods. Where consumer concern exists, such regulation will also provide the consumer with a basis for using irradiated foods with confidence.

12.1. Procedures and requirements for clearance for human consumption

Some countries have approached the procedure for regulation of irradiated foods by way of a general prohibition that can be removed when the requirements for such action are satisfactorily met. It is thus possible to control the irradiation of each and every food. The requirements to be met have included the points discussed in Section 11.1.

In the past, procedures for clearance of irradiated foods for human consumption have led to clearances either on an unconditional or a provisional basis for several food items and groups of related food products in different countries. These clearances were usually issued upon presentation of evidence of wholesomeness specific to the food in question.

The 1976 Joint FAO/IAEA/WHO Expert Committee on the Wholesomeness of Irradiated Food stressed that the microbiological, nutritional and toxicological approaches to the assessment of the wholesomeness of irradiated food must be based on the concept of food irradiation as a physical process, a concept which is gaining acceptance. Under this new concept it is possible to issue a regulation for the clearance of irradiated food on a much broader basis.

As part of the regulation of the commercial use of food irradiation, governments would like to be able to establish whether or not a particular food has been irradiated and with what dose. This interest has led to the study of various procedures for detecting and measuring physical, chemical and biological changes in foods.

Proteins, lipids and carbohydrates have been examined for characteristic chemical and physical changes. Techniques for detecting free radicals, such as electron spin resonance, have been applied. Histological methods have been used for examining the microstructure of foods. The biological function of living foods, such as potatoes, has been studied. The surviving microfloras have been differentiated from the normal ones that occur without irradiation.

Despite these and other investigations, no satisfactory method for identifying food as irradiated has been developed. While certain effects can be identified, there is not sufficient precision for regulatory purposes. Control of commercial food irradiation can therefore only be performed in the irradiation plant.

12.2. Regulations for control of the irradiation process, irradiated foods and their trade

The 1976 Joint FAO/IAEA/WHO Expert Committee on the Wholesomeness of Irradiated Food recommended that the irradiation of five food items be given unconditional clearance (wheat, potato, chicken, papaya and strawberries), and that provisional clearance should be given to three others: onion, cod and red fish, and rice. After this recommendation, the Codex Committee on Food Additives, which already in 1969 included the irradiation process within its terms of reference, developed, in accordance with the Codex Procedure for the Elaboration of World-Wide Codex Standards, a General Standard for Irradiated Foods. At its 13th Session, in December 1979, the FAO/WHO Codex Alimentarius Commission adopted a Recommended International General Standard for Irradiated Foods and a Recommended International Code of Practice for the Operation of Radiation Facilities Used for the Treatment of Foods.

The General Standard refers only to what relates to the irradiation process. It includes general requirements that apply to all irradiated foods and requirements applicable to the irradiation process. It also includes 'dose limits' for the maximum amount of radiation which may be absorbed by the individual food items listed above. These developments made it necessary to develop a legal framework that could serve as the basis for harmonization of national legislation and regulatory procedures that will enhance confidence among trading nations that food irradiated in one country and offered for sale in another has been subjected to commonly acceptable standards of wholesomeness, hygienic practice and irradiation-treatment control. Consequently, the International Atomic Energy Agency published in 1979 a document, *International Acceptance of Irradiated Food - Legal Aspects,* which contained Model Regulations for the Control of and Trade in Irradiated Foods. These Model Regulations provide valuable guidelines for Governments wishing to harmonize their national legislation relating to the practical application of food irradiation in accordance with the Codex Standard and Code of Practice. The main purpose of the Model Regulations is to establish and organize the appropriate measures for the three successive stages in the control of the irradiation of foods:

(1)   Approval of irradiation plants: the radiation treatment of foods is to be carried out in facilities licensed and registered for this purpose by the competent national authority.

(2)   Control of the food irradiation process: This ensures that irradiation processing is carried out by qualified persons and in accordance with the general requirements of the Codex Standard and Code of Practice. Plant managers and operators shall keep adequate records including quantitative dosimetry. Premises and records shall be open to inspection by appropriate authorities.

(3)   Control of trade in irradiated foods: This control is designed to prevent the entry into national and international trade of irradiated foods which have not been treated in conformity with provisions made in the Codex Standard and Code of Practice. In assuring comparability of control, certain general conditions are laid down concerning the labelling of irradiated foods, presentation and packaging which are considered essential to facilitate free movement of irradiated foods in international trade.

12.3.   Recent developments on evaluation of the wholesomeness and standardization of irradiated food

As of 1976 a large amount of data has been generated on the wholesomeness of irradiated foods and food components.

As already mentioned, early work largely consisted in testing individual foods by animal-feeding studies. More recent work has been directed to the greater use of the knowledge of the radiation chemistry of foods, the evaluation of irradiated complete diets, and treatment of irradiation as a general food process instead of considering radiation as a food additive.

The sum of valid completed investigations of wholesomeness has not demonstrated positive evidence of a toxicological or microbiological hazard associated with the consumption of any irradiated food. The normal nutritive values of the macronutrients of foods (protein, lipid and carbohydrate) are maintained. In some uses of irradiation, certain vitamins are labile to radiation, but the losses are either small or comparable to those associated with other commonly used food processes. A number of food-contacting packaging materials of all needed types have been found to cause no health hazard when irradiated.

Based on these data, the 1980 Joint FAO/IAEA/WHO Expert Committee on the Wholesomeness of Irradiated Food developed a recommendation on the acceptability of food irradiated up to an overall average dose of 10 kGy (1 Mrad). No toxicological hazard is caused by irradiating any food up to this dose in order to achieve various desirable objectives of food preservation, and hence foods treated in this way no longer need to be tested for toxicity. The favourable conclusions of the 1980 Expert Committee allowing the general use of the irradiation process up to an overall average dose of 10 kGy called for a revision of the Recommended Codex Standard for Irradiated Foods.

The Codex Committee on Food Additives therefore agreed at its 14th Session, held in November 1980, that the Codex Alimentarius Commission should be requested to initiate the Procedure for the Amendment of Codex Standards for the Recommended International General Standard for Irradiated Foods. The revision of the present Standard for Irradiated Foods is now being considered.

13. FOOD IRRADIATION FACILITIES

13.1. Radiation characteristics

The commonly used radionuclide and machine sources of ionizing radiation in food irradiation are:

(a) Mono-energetic γ-radiation from the radionuclides ^{60}Co or ^{137}Cs with 1.25 (average) and 0.66 MeV, respectively.
(b) Bremsstrahlung or X-radiation from accelerators (up to 5 MeV).
(c) Electron beams from accelerators (up to 10 MeV).

Only the first two of these types of radiation may be used for irradiation of relatively thick products (thicker than 7 cm) because of their greater penetrating

112

TABLE XXXIII. CHARACTERISTICS OF ^{60}Co AND ^{137}Cs

	^{60}Co	^{137}Cs
Typical source form	Metal	CsCl pellets
Half-life	5.3 years	30 years
Available specific activity	1 to 400 Ci/g[a]	1 to 25 Ci/g[a]
Gamma energy	1.17, 1.33 MeV	0.66 MeV
Power	65 Ci/W[a]	207 Ci/W[a]

Reprinted from Chemical and Food Applications of Radiation, Nuclear Engineering, Part XIX, Symposium Series 83, Vol. 64 (1968) 18.

[a] 1 Ci = 3.7 × 10^{10} Bq.

power. For shallow penetration and rapid conveyor speeds (as with grain irradiation), on the other hand, electron beams may provide a more uniform dose distribution at lower cost per unit mass of product when large amounts of product are involved.

13.1.1. Mono-energetic γ-rays

Mono-energetic γ-rays for use in food processing are usually obtained from large radionuclide sources containing either ^{60}Co or ^{137}Cs. The radiation characteristics of these radionuclides are given in Table XXXIII. ^{60}Co sources are made of the metal in various sizes and shapes, including rods and strips, while ^{137}Cs is usually supplied as CsCl pellets. To avoid contamination of the environment, both radionuclide sources are doubly encapsulated in stainless steel. To provide a source for a particular installation, it is assembled into an appropriate configuration such as a plaque or cylinder made up of pencils or rods of the radionuclides. By incorporating the requisite number of rods or pencils, a source of desired radiation output is secured. Narrow beams of γ-rays are attenuated exponentially in absorbing media. It is thus possible to compute the dose distribution in food samples irradiated even when very complicated source geometries, such as those with extended plaque sources, are used. The resulting depth-dose distribution in the food products closely resembles an exponential curve. A two-sided (bilateral) irradiation, obtained either by turning the sample or by having simultaneous irradiation from two opposite sides, would flatten the dose distribution, having the effect of reducing the ratio of maximum to minimum dose within the product (Fig.30).

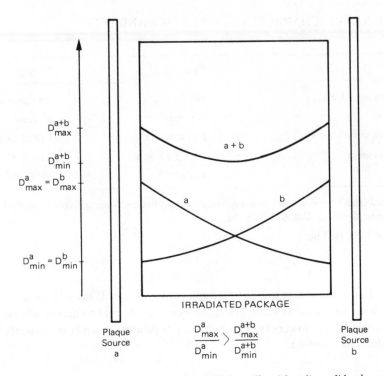

FIG.30. Depth-dose curves in a package irradiated bilaterally with radionuclide plaque sources a and b. Curve a is depth dose for plaque source a only, curve b is that for plaque source b only, and curve a + b is the combined depth-dose curve.

13.1.2. Bremsstrahlung and X-rays

In contrast to radionuclide sources, bremsstrahlung and X-ray irradiators emit a broad-band spectrum consisting of photons of numerous energies (Fig.31).

Since the attenuation of photons is dependent on their energy, i.e. the lower-energy photons are more rapidly attenuated, the composite depth-dose curve can be considered as a superposition of individual depth-dose curves of equally-spaced incremental energy intervals over the entire spectrum.

13.1.3. Electrons

Electrons emitted by accelerators (Van de Graaff, Insulating Core Transformer, Dynamitron, LINAC) have fairly narrow spectral energy limits. The range of an electron in an absorbing medium is limited and is closely related to its energy (0.5 cm of water per MeV). 10-MeV electrons can be used to perform one-sided irradiations of 3.9-cm-thick targets (water equivalent) with dose-uniformity ratios

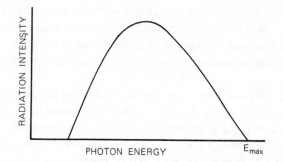

FIG.31. *Typical shape of the photon spectrum from a bremsstrahlung source of several MeV maximum energy, where* E_{max} *is the highest photon energy, equal to the maximum electron energy.*

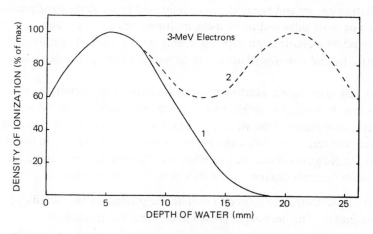

FIG.32. *Density of ionization at various depths in water irradiated with 3-MeV electrons. (1) Irradiated from one side; (2) Irradiated from two opposite sides.*

approaching 1.3, whereas 3-MeV electrons can only be used for thin-layer treatment since they penetrate to a depth of only 1.5 cm in water (see Fig.32).

If the product package to be treated is much thicker than the range of the irradiating electrons, only a surface treatment is possible. Two-sided through-irradiations to reduce the uniformity ratio can be undertaken when the electron range is greater than the thickness of the product package.

13.2. Dose distribution in the product

The applied dose should not be higher or lower than is needed to achieve the desired effect. Finding and applying the appropriate dose level is the key to

115

the technologically and economically proper application of the irradiation process to food. As stipulated in Section 12.1, effective dose control can only be performed in the irradiation plant. A major factor in the design of an irradiation facility is the uniformity of the distribution of absorbed dose in the given product.

Dose uniformity is a composite state depending on the distribution of dose both through the depth of the product and laterally within it.

The depth-dose is the distribution along the centre-line of the product, perpendicular to the plane of the source in the case of radionuclide plaque sources, or parallel to the direction of the beam axis in accelerator irradiation. The lateral dose distribution is the distribution of dose within the product in a plane taken parallel to the plane of the radionuclide plaque source and in a plane taken perpendicular to the direction of the beam in accelerator irradiation.

The depth-dose uniformity is limited by product density, product thickness, and radiation energy and type; it can be improved by irradiating the product from two or more sides and/or by using multipass irradiator systems.

Lateral dose distribution depends mainly on the source-to-product geometry. Lateral dose distribution can be improved in the following ways:

(a) By introducing additional source material in the vicinity of the low-dose product regions. In electron accelerators, a similar effect is achieved by using scatter plates in the vicinity of low-dose regions or by electronically modifying the scan speed (allowing the beam to scan more slowly in low-dose regions). In X-ray machines such an effect can be obtained by varying the different partial beam currents (increasing the flux in the low-dose regions).

(b) By allowing the source to overlap the product in the vicinity of the low-dose points. This technique is used in radionuclide irradiators.

(c) By allowing the product to move or sweep past the source at a uniform speed or in a shuffle-dwell motion (i.e. the product moves discontinuously past the irradiation source, with periods of movement and stationary positions alternating). These techniques are used in both radionuclide and accelerator irradiators and contribute to improving the uniformity of dose along any line in the product that is parallel to the direction of its motion.

The maximum and minimum absorbed doses of radiation throughout a container, package or layer of food depend on the depth-dose and lateral-dose distributions in that product. The magnitudes of the maximum and minimum doses consequently determine the dose uniformity ratio. For detailed information on dose distribution patterns and measurement in foods treated with ionizing radiation from different irradiator designs the reader is referred to Chapter III of the Manual of Food Irradiation Dosimetry (Technical Reports Series No. 178, IAEA, Vienna (1977)).

116

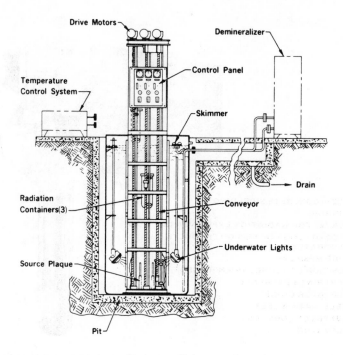

FIG. 33. Water-shielded ^{60}Co research station.

13.3. Radionuclide source irradiators for research and pilot-plant operations

Some 60 facilities for research on food irradiation have been built and installed in government, university and private laboratories in many countries all over the world. Most such facilities use ^{60}Co. A few have ^{137}Cs and others have machine sources of radiation.

The shielding problem of radionuclide sources has been solved in two principal ways: by the use of a sufficient depth of water and by the use of a solid absorber such as concrete or lead.

Figure 33 is a schematic diagram of a water-shielded ^{60}Co research irradiator. In this type 1.18 PBq (32 000 Ci) of ^{60}Co are employed. The products for irradiation, up to 35 × 45 × 15 cm in size, are placed in water-tight containers and lowered to the source, which is kept at the bottom of the pool.

Figure 34 is an example of a source in which the ^{60}Co is raised for use. Since it is not necessary to submerge the samples in the water, this arrangement is more convenient. While the source is at the bottom of the pool, samples for irradiation are taken into a cave made of shielding material (e.g. concrete) which is built over the pool. The samples are located at a specific and appropriate

117

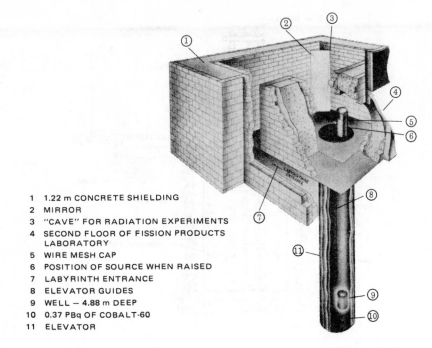

1   1.22 m CONCRETE SHIELDING
2   MIRROR
3   "CAVE" FOR RADIATION EXPERIMENTS
4   SECOND FLOOR OF FISSION PRODUCTS
     LABORATORY
5   WIRE MESH CAP
6   POSITION OF SOURCE WHEN RAISED
7   LABYRINTH ENTRANCE
8   ELEVATOR GUIDES
9   WELL – 4.88 m DEEP
10 0.37 PBq OF COBALT-60
11 ELEVATOR

FIG.34. *University of Michigan* ^{60}*Co irradiation cave. Reprinted from CHARLESBY, A.,
Radiation Sources, Pergamon Press, Oxford (1964).*

distance from the source. Personnel then leave the cave, and the source is
brought out of the water by remote operation. On completion of the irradiation
it is lowered into the water.

    To provide for a continuous operation, some facilities have included a
carrier system which transports product from outside the shielding to the source
and back outside again. Figure 35 shows such a facility schematically.

    A simple self-contained unit is shown in Fig.36. The radiation source may
be made of as many as twelve ^{60}Co pencils. The shielding is steel-encased lead
and can accommodate 1.85 PBq (50 000 Ci) of ^{60}Co. Four product-carrier
boxes (19 X 19 X 50.8 cm) rotate round the ^{60}Co source and provide a useful
irradiation volume of 81 ltr. For an absorber of density 0.5 and a loading of
1.85 PBq (50 000 Ci) of ^{60}Co, the dose rate is 9 Gy (900 rad) per minute.

    Some mobile irradiators containing either ^{60}Co or ^{137}Cs have also been built.

13.4. Radionuclide source irradiators for practical application

    A production facility should be operated with optimum efficiency for
utilization of radiation energy. It is difficult to generalize when dealing with

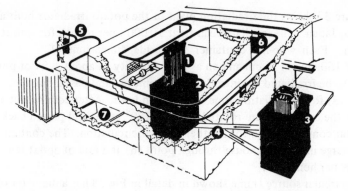

FIG.35. Gamma cell: (1) Cobalt source (elev.); (2) Inner pool; (3) Cask pool; (4) Transfer tubes; (5) Overhead conveyor; (6) Trays; (7) Man-trap (Courtesy US Army Food Radiation Lab., Natick, Mass., USA)

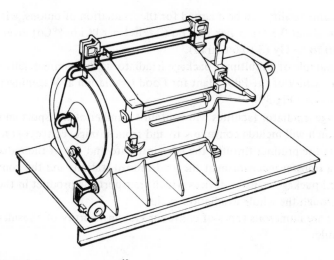

FIG.36. Self-contained lead-shielded ^{60}Co research irradiator (courtesy Atomic Energy of Canada Ltd, Ottawa).

food irradiation, since some facilities will be designed to handle only one food item, whereas others will be designed to handle many different food items.

Some examples of large-scale radionuclide source irradiators are described below.

For the irradiation of potatoes, onions and other root products, the containers are usually bulky, having a capacity of 1 t or more in a single unit. To co-ordinate with the methods of harvesting, storage and shipment of the products, it is best to use the same container within the radiation facility.

119

Figure 37 shows the irradiation plant of the potato irradiator built at Hokkaido, Japan. The irradiator can treat 10 000 t per month for sprout inhibition. Each wire-net container has inside dimensions of 100 cm $\times$ 160 cm $\times$ 130 cm and an average capacity of about 1.5 t of potatoes. Potatoes to be irradiated are introduced at point (1) and pass along the conveyor (3) for a circulation time of about an hour. After coming out at point (4), the containers are rotated through 180° at (5) and passed back onto the circular conveyor again for another hour of circulation. The containers finally emerge in the warehouse through point (6) at a rate of about ten containers per hour.

The circular source frame shown in detail in Fig.38 has a diameter of 100 cm and contains 36 ^{60}Co source rods, each 100 cm long and about 280 TBq (7500 Ci) activity. The average values of D_{min} and D_{max} are measured as 60 and 150 Gy (6 and 15 krad) respectively, i.e. the dose uniformity ratio is 2.5.

The same facility can be adapted for the irradiation of onions, with a capacity of about 0.5 t·h^{-1} per 10 TBq (2 t·h^{-1} per kCi of ^{60}Co), over a dose range of 30 to 66 Gy (3 to 6.6 krad).

An example of a multipass package irradiator is the commercially used ^{60}Co γ-ray facility of the Pilot Plant for Food Irradiation at Wageningen, Netherlands (see Fig.39).

Package irradiator facilities will contain a number of transport and transfer systems which will include conveyors to and from storage, conveyors for transferring the product through the labyrinth into and out of the radiation chamber, a transfer mechanism at the load/unload stations and the conveyor system, and package transfer devices which transport the product in the designed fashion through the whole irradiation cycle.

There are numerous types of conveyor systems worthy of consideration. They include:

Cartons stacked in hanging containers moving along monorails or similar supporting beams;
Roller conveyors with mechanical pushing devices to transport cartons;
Cartons in trays that slide or move on wheels with the aid of mechanical or pneumatic transport devices;
Cartons moved by electrically driven carts.

The cartons are usually irradiated on both sides to flatten the dose distribution in the direction perpendicular to the source plaque. To attain improved uniformity of lateral-dose distributions, complex movement schemes are often followed during a complete irradiation cycle. It is usually necessary to provide for multiple presentations of the target material to the source.

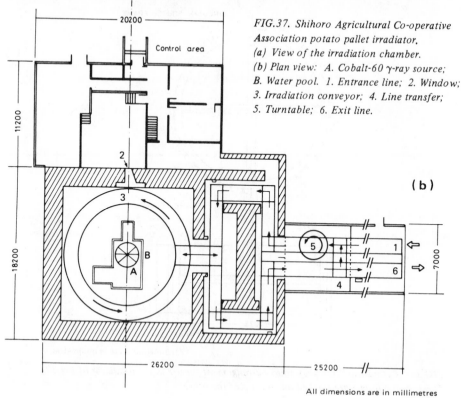

FIG.37. *Shihoro Agricultural Co-operative
Association potato pallet irradiator.*
(a) View of the irradiation chamber.
(b) Plan view: A. Cobalt-60 γ-ray source;
B. Water pool. 1. Entrance line; 2. Window;
3. Irradiation conveyor; 4. Line transfer;
5. Turntable; 6. Exit line.

All dimensions are in millimetres

CIRCULAR SOURCE FRAME

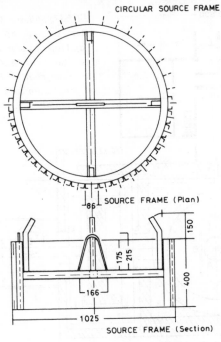

SOURCE FRAME (Plan)

SOURCE FRAME (Section)

All dimensions are in millimetres

FIG.38. Cylindrical ^{60}*Co γ-ray source arrangement for use in the irradiator of Fig.37.*
Three source frames are stacked one on top of the other.

122

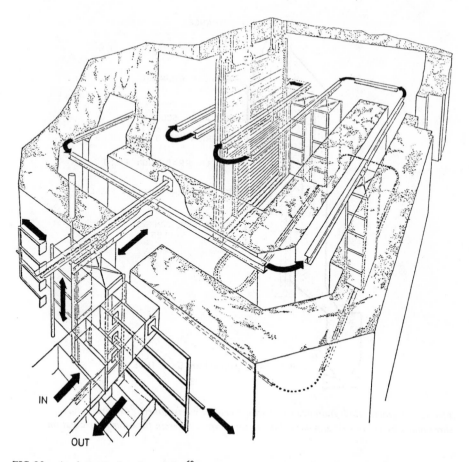

FIG.39. *A schematic drawing of the ^{60}Co gamma irradiation pilot-plant at Wageningen.*
It is a two direction, four-pass automatic irradiator which gives two-sided irradiation. Five
boxes, each 50 cm $\times$ 33 cm $\times$ 27 cm are carried in a container hanging from an overhead monorail.
Vertical transfer of the product occurs outside the irradiation chamber. The plaque source can be
divided into two parts for low-dose irradiation of potatoes and onions. With a 215 kCi (8 PBq)
source (Oct. 1977), a throughput of 3.5 $\times$ 10^5 krad·kg·h^{-1} (3.5 $\times$ 10^3 kGy·kg·h^{-1} $\simeq$ 0.97 kW)
at a bulk density of 0.5 g/cm^3 (500 kg/m^3) and a uniformity ratio of 1.3 was obtained.

A typical design for a continuous-flow irradiator for disinfestation of
grains is illustrated in Fig.40. Grains are transferred through pneumatic feed
ducts and irradiated as they drop through annular zones surrounding the source.
The irradiation chamber is divided laterally into three concentric annular zones,
with different retention times in each zone. In this irradiator, when providing
a typical throughput, the control valves are kept closed for periods of 130 s
for the interior zone, 195 s for the middle zone, and 280 s for the exterior zone,
thus compensating for the lower dose rates in the outer zones.

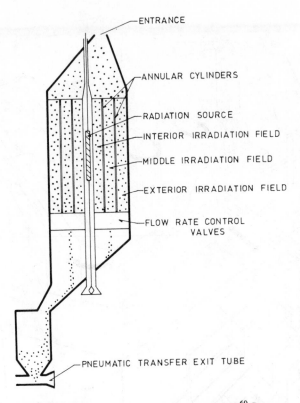

ENTRANCE

ANNULAR CYLINDERS

RADIATION SOURCE

INTERIOR IRRADIATION FIELD

MIDDLE IRRADIATION FIELD

EXTERIOR IRRADIATION FIELD

FLOW RATE CONTROL VALVES

PNEUMATIC TRANSFER EXIT TUBE

FIG.40.   A gravity-flow grain irradiator with a centrally positioned ^{60}Co γ-ray source, surrounded by concentric annular cylinders for controlling grain flow during irradiation.

An effort was also made to decrease the uniformity ratio and increase the radiation efficiency by mounting a series of baffles within the annular zones, thus causing more turbulent grain flow. By adjusting the individual retention times in each zone, a uniformity ratio of about 2 has been achieved.

13.5. Accelerator irradiators for research and practical applications

The primary radiations in various types of accelerators (also called machine-source irradiators) are electrons accelerated to high energies. Other accelerated particles are not used in food irradiation. In electron irradiators, the charged-particle beam emerging from the accelerator, i.e. the electron beam, can be used directly for treatment of the sample. Alternatively, this electron beam can be converted into X-rays (also called bremsstrahlung) by being made to strike a conversion target. The electrons are stopped in a high-atomic-number absorber and the kinetic energy of the decelerated electrons emerges in the form of bremsstrahlung from the converter.

Normally, accelerator irradiators utilize continuously moving conveyors to achieve a more uniform lateral-dose distribution in the direction of product motion.

13.5.1. High-energy electron sources

The high-energy electron sources are primarily beam generators. All these generators utilize the negative electric charge of the electron to impart energy to the electron through the application of an accelerating voltage. Except in the case of the linear accelerator, the full direct-current accelerating potential is developed. This limits the practical potentials to below about 4 MV. The linear accelerator, not having this limitation, is useful above 4 MV. Principal parts of each generator are the electron source and acceleration tube. Electrons are obtained from a hot wire (or cathode) and are subjected to the accelerating potential within the confines of an envelope or tube providing a vacuum. They emerge into air after penetrating a suitable tube window. The following machine sources are considered:

(a)   Van de Graaff accelerator

This accelerator is portrayed schematically in Fig.41. Within a pressurized housing, a corona discharge is directed onto an endless belt of non-conducting material. The resultant 'charge' is conveyed physically to the upper part of the housing and transferred to an isolated conductive shell. By accumulating sufficient 'charge' this shell develops a high potential, which is applied to the acceleration tube and to the electrons. The Van de Graaff accelerator delivers a constant current of monoenergetic electrons.

(b)   Insulating core transformer

The belt-driven Van de Graaff accelerator has been made largely obsolete by the higher-powered insulating core transformer. The power supply for this unit consists of a three-phase transformer with multiple secondaries, each of which is insulated from the other. The alternating current in each secondary is rectified, and the individual direct current outputs are connected in series to provide the high voltage for acceleration of the electrons. The transformer secondaries are made into modules or decks. Each deck contains the secondary coil, rectifier and other circuit components and has an output voltage of approximately 50 kV. The decks are stacked vertically and connected in series. The number of such decks determines the output voltage.

125

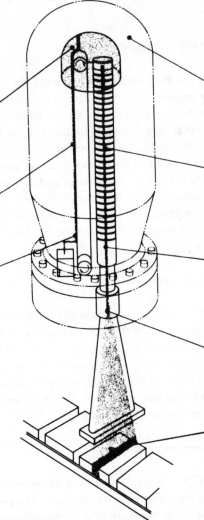

3

At the terminal, the charge is automatically transferred from the belt to the terminal, thereby establishing a high potential of voltage difference with respect to the lower end of the accelerator.

2

The belt mechanically carries the charge to an insulated, hemispherical, high-voltage terminal.

1

Electric charge is sprayed on a rapidly moving insulating belt.

4

The high-voltage terminal is insulated from the shell of the accelerator by an atmosphere of compressed nitrogen, which prevents arc-over.

5

A glass and metal tube, maintained at a very high vacuum, provides the only path for the electrons to escape from the cathode.

6

The electrons forming the high-energy beam are accelerated to extremely high velocities by the potential difference between the terminal and the lower end of the accelerator.

7

This electron beam is scanned by magnetic coils to cover uniformly the product passing beneath.

8

The dosage received by the irradiated product depends on the speed of the conveyor belt, the energy of the beam, and the width of scan.

FIG.41.   Van de Graaf Accelerator (courtesy High Voltage Engineering Corp., Burlington, Mass., USA).

126

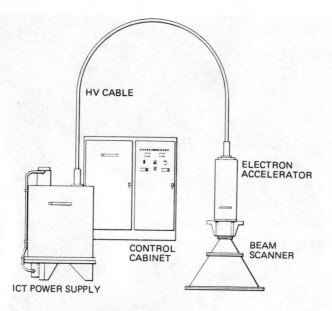

FIG.42. Major components of insulating core transformer electron beam generator (courtesy High Voltage Engineering Corp., Burlington, Mass., USA).

The terminal voltage of the power unit is connected to the cathode end of the acceleration tube. This is constructed of glass ring insulators placed between highly polished metal electrodes. A uniform voltage gradient along the tube is obtained through the use of resistors connecting each of the electrodes. The electron beam originates from a tungsten filament. It is focussed and the electrons are accelerated inside the evacuated tube to an energy corresponding to the output voltage of the power supply.

The monoenergetic electron beam emerges from the tube through a thin metal window. The beam is approximately circular in cross-section and can be from a fraction of a mm to several cm in diameter. It is most intense in the centre. To provide lateral-dose uniformity across the conveyor, the electron beam is magnetically deflected back and forth (100–300 Hz) inside an evacuated extension of the electron tube and emerges through a titanium-foil window. To secure the indicated uniformity, the accelerating potential from the power supply must be constant. With beam-scanning, the scan frequency and the conveyor speed must be co-ordinated so as to ensure multiple overlap of successive scan traces, thereby taking care that the whole of the product is completely irradiated. Figure 42 shows schematically the major components of the insulating core-transformer electron-beam unit.

FIG.43. Dynamitron Electron Accelerator; pressure vessel with mounted 'dees' (courtesy AEG-Telefunken, Wedel (Holstein), FRG).

(c) Dynamitron

The Dynamitron is a high-voltage DC electron accelerator. The voltage is generated by a system of cascaded solid-state rectifiers driven in parallel from an RF oscillator via 'dees' which surround the rectifier stack (see Fig.43). The electrodes are capacitively coupled through high-pressure SF_6 gas to corona rings which are connected to the rectifiers. The same RF potential is impressed on each rectifier and the DC components are added in series to establish the desired total voltage, which is impressed across the electron beam tube. Electrons generated by the filament are accelerated in vacuum in the beam tube to an

128

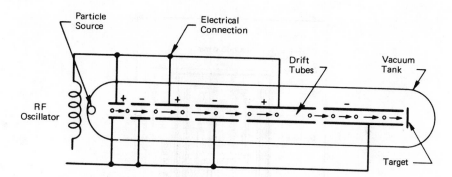

FIG.44. A linear accelerator. The separation between accelerating gaps, which is the distance traversed by the particles during one half cycle of the applied electric field, becomes greater as the velocity of the particle increases. At any instant adjacent electrodes carry opposite electric potentials. These are reversed each half cycle.

TABLE XXXIV. CALCULATED CONVERSION EFFICIENCIES FOR X-RAY PRODUCTION

Electron energy (MeV)	Conversion efficiency (%)	
	Aluminium	Tungsten
10.0	7.7	30
5.0	4.0	19
3.0	2.5	14
2.0	1.8	10
1.0	0.9	6
0.5	0.4	3

energy corresponding to the total voltage. The beam of electrons is magnetically scanned at high frequency in a scan horn to form a 'screen' of electrons, which leave through a thin window and emerge into the air. Voltages from 500 to 1500 kV are available.

(d)   Linear accelerator (LINAC)

This kind of accelerator exists in several forms. Electrons are given energy by properly phased sequential exposure to a given potential difference or,

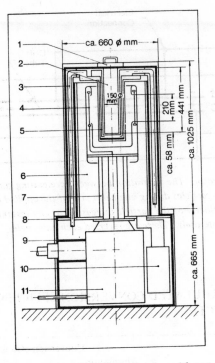

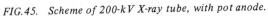

FIG.45. Scheme of 200-kV X-ray tube, with pot anode.

1. *Cover with safety contact*
2. *Irradiation chamber*
3. *Anode cooling water*
4. *Cathode*
5. *Double-walled pot anode*
6. *Vacuum*
7. *Oil-filled cathode insulator*
8. *Base (850 × 850 mm)*
9. *Lead shielding*
10. *Ion getter pump*
11. *Transformer for cathode heating
   with high-voltage cable and plug 200 kV*

(Courtesy AEG-Telefunken, Hamburg, FRG.)

alternatively, by keeping continuously in step with a moving electromagnetic field. This approach avoids the limitations imposed by the ability to handle the full accelerating potential. Consequently energies greater than 4 MeV can be given to electrons. (One form of this accelerator is shown schematically in Fig.44.) The essentially monoenergetic electron beam is pulsed. As with other electron-beam generators, the beam may be scanned over a given area. Here

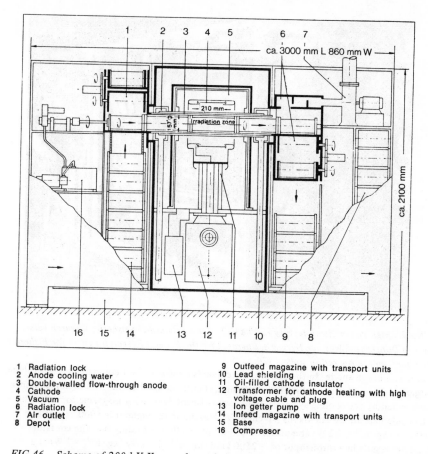

1 Radiation lock
2 Anode cooling water
3 Double-walled flow-through anode
4 Cathode
5 Vacuum
6 Radiation lock
7 Air outlet
8 Depot

9 Outfeed magazine with transport units
10 Lead shielding
11 Oil-filled cathode insulator
12 Transformer for cathode heating with high
   voltage cable and plug
13 Ion getter pump
14 Infeed magazine with transport units
15 Base
16 Compressor

FIG.46. Scheme of 200-kV X-ray tube with flow-through anode tube (courtesy AEG-Telefunken, Hamburg, FRG).

again, in a pulsed-beam accelerator, to ensure that the whole of the product is completely irradiated, the pulse repetition rate as well as the scanning frequency and the conveyor speed must be co-ordinated. Although linear accelerators can provide electrons with energies greater than 10 MeV, their use in food irradiation is limited to providing a maximum of 10 MeV (see Sections 1.4.5 and 11.10).

13.5.2. X-ray sources

When high-energy electrons strike a conversion target, X-rays are formed. The kinetic energy of the decelerated electrons is emitted as a broad-band spectrum consisting of photons of numerous energies (see Fig.31). The

FIG.47. High power X-ray facility (200 kV, 150 mA) aboard the fisheries research vessel 'Anton Dohrn' (ex-'Walther Herwig'), operated by the Federal Research Centre for Fisheries, Hamburg, and in operation since 1970.

The device has an anode. In the picture, the casing is removed. Left compartment: revolving magazine for loading the cylindrical containers. Centre compartment: irradiation chamber surrounded by the cooling system. Right compartment: revolving magazine for unloading. The containers are 20 cm dia. by 80 cm long. Maximum throughput is $8000 \text{ krad} \cdot \text{kg} \cdot \text{h}^{-1}$ ($80 \text{ kGy} \cdot \text{kg} \cdot \text{h}^{-1} \simeq 22 \text{ W}$) at a uniformity ratio of ~ 2.0; use of only half the container diameter results in a throughput of $\sim 2500 \text{ krad} \cdot \text{kg} \cdot \text{h}^{-1}$ ($25 \text{ kGy} \cdot \text{kg} \cdot \text{h}^{-1} \simeq 7 \text{ W}$) at a uniformity ratio of 1.4.

conversion efficiency of electrons to X-rays depends on the target material and the electron energy. This relationship for aluminium and tungsten is shown in Table XXXIV.

The relatively low efficiency for converting electrons to X-rays is not necessarily a deterrent in the use of this kind of source. The high penetrating power of X-rays is of great advantage in certain applications. The capability of selecting the photon spectrum of X-irradiation by using a particular level of energy for the electrons of the X-ray source and by using a particular tube target material affords a convenient method for controlling the penetrability of the X-radiation. This is also useful in minimizing shielding requirements for the source. As noted in Sections 1.4.5 and 11.10, the maximum energy level for X-rays used in food irradiation is 5 MeV.

Some X-ray irradiators are designed to give a very broad beam of electrons striking an extended converter. This converter may be as large as an isotope plaque source so as to achieve dose uniformity by source overlap.

Figure 45 shows schematically a stationary type of accelerator irradiator, e.g. the pot-anode X-ray tube system. In this system, the target product is placed inside the irradiation chamber (inside the pot) and is irradiated by bremsstrahlung from all sides simultaneously. Upon completion of the irradiation the accelerator is turned off, the irradiated product is removed, and a new batch inserted.

A flow-through X-ray facility is shown schematically in Fig.46. Figure 47 shows another such facility installed aboard a fishing vessel. The material to be irradiated can be moved through the irradiation chamber at an adjustable speed according to the dose required. The material is prepacked in standard containers for transport through the irradiator.

14. COMMERCIAL ASPECTS OF FOOD IRRADIATION

Regardless of how business is conducted in a given country, if irradiation is to be used as a food preservation or improvement process, it must serve a useful purpose. Particular ways in which irradiation can be useful have been described in Section 7. How these methods are made to fit into a country's food preservation and distribution system will vary greatly with local conditions, and it is difficult to give specific information on how it is to be done. Certain steps with general application can, however, be identified:

(a) Development of the technology of product and process.
(b) Government approval for the process and packaging materials, if applicable, including regulations for them.
(c) Pilot-scale production to develop a process suitable for commercial scale and to test the acceptance of the product in the market place. This will include determination of cost and determination of cost acceptance by consumer.
(d) Assuming a favourable situation as determined in (c), production and distribution are increased to supply the irradiated food under the best attainable economies in competition with other foods on a broad market basis.

The consumer is the ultimate judge of the value of the product. His willingness to use it and to pay for it determines the commercial success of a product. In some cases, irradiated foods will find justification in better quality or reduction of a health hazard; in others, there will be no product improvement, but the advantage will be one of reduced cost. One or the other, or a combination, will form the basis of the consumer's use of the irradiated food.

14.1. Consumer acceptance of irradiated foods

At least three areas of consumer acceptance of irradiated foods can be identified:

(a)   The broad one of concern about the fact that the foods have been treated by radiation;

(b)   The acceptance of those foods which are identical in quality with unirradiated counterparts as far as the consumer can determine;

(c)   The acceptance of those foods which as a consequence of irradiation have some kind of quality improvement.

The first item has several facets. There is the basic one of the consumer's actual acceptance of irradiation. The food manufacturers and distributors are also concerned about the consumer's reaction and the risks they take in the manufacture and sale of irradiated foods owing to the uncertainty of consumer acceptance. Finally, there are the government regulatory agencies, whose judgement of the safety of irradiated foods for human consumption allows such foods to be distributed to the public. Their responsibility in arriving at this judgement is a weighty one and they must be able to convince the consuming public that there is no hazard.

Many irradiated foods will not change their character as a consequence of being irradiated, and the consumer will see no difference from what he has been accustomed to. Foods irradiated for insect-infestation control or delay of senescence will have normal characteristics. These the consumer should accept without difficulty, provided the first aspect presents no problem and provided the cost is satisfactory.

For irradiated foods with improvement in quality there should be increased consumer acceptance. Just how much extra cost can be justified will vary with many circumstances. It is likely that answers to this question can be obtained only when irradiated foods are distributed in competition with other foods.

Concern has been expressed about damage to product quality by radiation. As noted earlier, this can happen. For some foods this will so reduce consumer acceptance as to prevent application of irradiation to such foods. For others there is no detectable difference, and there will be no quality problem. For still others there is a detectable difference in quality, and the question is: how much change will be tolerated? This is especially relevant if viewed in terms of an advantage gained (e.g. availability through shelf-life extension).

Some experience of consumer response to irradiated foods has been gained, while true commercial production and distribution of irradiated foods has been limited. Potatoes, irradiated to inhibit sprouting, have been sold commercially and have been well received. There have been numerous market tests on a limited scale in various countries as well as other tests in which consumer response to irradiated foods has been measured (see, for example, Tables XXI and XXII).

All such experiences encourage the opinion that there can be good consumer acceptance of irradiated foods.

14.2. Economics of irradiation

The cost elements for irradiation are similar to those for any food manufacturing and distribution operation: (a) fixed costs associated with the facility, and (b) variable costs associated with the operation of the facility and with the transport of the food from point of production to point of consumption, including storage. Variable costs are normally charged as they occur. They include, for example, labour and supervision, utilities, supplies, maintenance and repair, and taxes. Fixed costs are usually prorated on some basis of time, which is related to the presumed useful life of the facility. This proration may be somewhat arbitrary and may be employed to permit recovery of the capital cost before the end of the useful life of the facility.

Both variable and fixed costs are prorated over the number of units produced and thus the cost per product unit is obtained.

The key factor in capital costs is the size of the radiation source, which itself is a major cost item. Size of source also determines other costs, such as that of radiation-shielding structures, and of radiation source strength maintenance. Source size is determined by three principal factors: dose requirement, product throughput requirement, and source efficiency.

The kind of radiation source employed can affect costs. There are at least two possible γ-sources (e.g. ^{60}Co, ^{137}Cs), both of which seem to have roughly equal overall costs. There is also the possibility of using X-ray machines, the economics of which are not fully available. It appears quite certain, however, that electron beams, which have a substantially higher conversion factor for line-power to ionizing radiation than X-ray units, cost less than X-rays. For large installations, it seems probable that the use of electron beams is the most economical form of ionizing radiation. However, the low penetrating power of electron beams has to be taken into consideration when deciding which type of source to employ for food processing.

Dose requirement varies from about 50 Gy to 50 kGy (5 krad to 5 Mrad), i.e. by a factor of 1000, depending on the particular application of the irradiation. Product throughput is determined by the hourly rate or capacity and the hours of operation. Increasing the hours of operation reduces the source size requirement. The greater the source efficiency the smaller is the source size requirement. Source efficiency is related to source type and design and to product-handling. In general, the greater the range between the maximum and minimum dose given to a product the greater is the efficiency of the source. Narrow limits yield small efficiencies.

In a properly built and operated plant, labour requirements, and therefore labour costs, are not likely to be large, primarily owing to the simple operation involved.

A comparison of both fixed and variable costs for low- and medium-dose applications tends to approximate those of other comparable processes. The costs for high-dose applications (e.g. radappertization) may vary either way, larger or smaller than those for thermal canning. Larger costs may be justified by improved product quality.

15. LITERATURE SOURCES

15.1. General

Much of the published work on food irradiation exists in the standard publications of food science and technology and related fields. Many reports have been issued by governments and official groups which, for various reasons, have now ceased providing these publications. To the extent that they are still available, they can be valuable sources of information on food irradiation. Current publishers in the field of food irradiation are:

International Atomic Energy Agency, Vienna.

World Health Organization, Geneva.

Joint FAO/WHO Food Standards Programme, Codex Alimentarius Commission, Food and Agriculture Organization of the United Nations (FAO), World Health Organization (WHO), Rome.

International Facility for Food Irradiation Technology (IFFIT), Wageningen, Netherlands.

Bibliography in Irradiation of Foods (Bibliographie zur Bestrahlung von Lebensmitteln), Bundesforschungsanstalt für Ernährung, Karlsruhe, Federal Republic of Germany.

International Project in the Field of Food Irradiation (IFIP), Karlsruhe, Federal Republic of Germany.

United States Army Natick Research and Development Command, Natick, Massachusetts, USA.

Association Internationale d'Irradiation Industrielle (AIII), Lyon, France.

Food Irradiation Japan, Japanese Research Association for Food Irradiation, Tokyo, Japan.

Radiotraitements (Radiation Processing), Association pour la Promotion Industrie-Agriculture (APRIA), Paris, France.

STH Berichte, Institut für Strahlenhygiene des Bundesgesundheitsamtes, Berlin (West).

15.2.   FAO/IAEA publications in the field of food irradiation[1]

Radiation Control of Salmonellae in Food and Feed Products
Technical Reports Series No.22 (1963) STI/DOC/10/22

(148 pp., Austr. Sch. 115)

Food Irradiation
Proceedings of a Symposium (1966) STI/PUB/127

(957 pp., Austr. Sch. 660)

Application of Food Irradiation in Developing Countries
Technical Reports Series No.54 (1966) STI/DOC/10/54

(183 pp., Austr. Sch. 144)

Microbiological Problems in Food Preservation by Irradiation
Panel Proceedings Series (1967)   STI/PUB/168

(148 pp., Austr. Sch. 100)

Preservation of Fruit and Vegetables by Radiation
Panel Proceedings Series (1968) STI/PUB/149

(152 pp., Austr. Sch. 120)

Elimination of Harmful Organisms from Food and Feed by Irradiation
Panel Proceedings Series (1968) STI/PUB/200

(118 pp., Austr. Sch. 100)

Enzymological Aspects of Food Irradiation
Panel Proceedings Series (1969) STI/PUB/216

(110 pp., Austr. Sch. 100)

---

[1] Orders may be addressed to:
Publishing Section, International Atomic Energy Agency,
Vienna International Centre, P.O. Box 100,
A-1400 Vienna, Austria

Microbiological Specifications and Testing Methods for Irradiated Food
TechnicalReports Series No.104 (1970) STI/DOC/10/104
(121 pp., Austr. Sch. 136)

Preservation of Fish by Irradiation
Panel Proceedings Series (1970) STI/PUB/ 196
(176 pp., Austr. Sch. 160)

Radurization of Scampi, Shrimp and Cod
TechnicalReports Series No.124 (1971) STI/DOC/10/124
(93 pp., Austr. Sch. 116)

Disinfestation of Fruit by Irradiation
Panel Proceedings Series (1971) STI/PUB/299
(180 pp., Austr. Sch. 160)

Report of a Consultation Group on the Legal Aspects of Food Irradiation
(1973) STI/DOC/59 (32 pp.)

Factors Influencing the Economical Application of Food Irradiation
Panel Proceedings Series (1973) STI/PUB/331
(137 pp., Austr. Sch. 140)

Radiation Preservation of Food
Proceedings of a Symposium (1973) STI/PUB/317
(774 pp., Austr. Sch. 760)

Aspects of the Introduction of Food Irradiation in Developing Countries
Panel Proceedings Series (1973) STI/PUB/362
(113 pp., Austr. Sch. 120)

Improvement of Food Quality by Irradiation
Panel Proceedings Series (1974) STI/PUB/370
(188 pp., Austr. Sch. 180)

Requirements for the Irradiation of Food on a Commercial Scale
Panel Proceedings Series (1975) STI/PUB/394
(219 pp., Austr. Sch. 250)

Manual of Food Irradiation Dosimetry
Technical Reports Series No.178 (1977) STI/DOC/10/178
(161 pp., Austr. Sch. 240)

Food Preservation by Irradiation
Proceedings of a Symposium (1978) STI/PUB/470
(Vol. I., 595 pp., Austr. Sch. 760)
(Vol. II., 429 pp., Austr. Sch. 550)

Decontamination of Animal Feeds by Irradiation
Proceedings of an Advisory Group Meeting (1979) STI/PUB/508

(153 pp., Austr. Sch. 250)

International Acceptance of Irradiated Food: Legal Aspects
Legal Series No.11 (1979) STI/PUB/530

(70 pp., Austr. Sch. 120)

Combination Processes in Food Irradiation
Proceedings of a Symposium (1981) STI/PUB/568

(467 pp., Austr. Sch. 720)

15.3. Food Irradiation Newsletter[2]

Joint FAO/IAEA Division of Isotope and Radiation Applications of Atomic Energy for Food and Agricultural Development; published periodically as of March 1977.

15.4. Food Irradiation Information

Published periodically as of November 1972 by the International Project in the Field of Food Irradiation (IFIP), Karlsruhe, Federal Republic of Germany.

15.5. Selection of books, reports, etc., published since 1970

WORLD HEALTH ORGANIZATION, Wholesomeness of Irradiated Food with Special Reference to Wheat, Potatoes and Onions (Report of a joint FAO/IAEA/WHO Expert Committee), Technical Report Series No.451, WHO, Geneva (1970).

TILTON, E.W., COGBURN, R.R., BROWER, J.H., Critical evaluation of an operational bulk-grain and packed-product irradiator, Int. J. Radiat. Eng. 1 (1971) 49—59.

DRAGANIC, I.G., DRAGANIC, Z.D., Radiation Chemistry of Water, Academic Press, New York and London (1971).

INTERNATIONAL COMMISSION ON RADIATION UNITS AND MEASURE-MENTS, Radiation Quantities and Units, ICRU Report 19 and Supplement, ICRU, Washington, DC (1971).

---

[2] Orders may be addressed to:
Food Preservation Section, International Atomic Energy Agency,
Vienna International Centre, P.O. Box 100,
A-1400 Vienna, Austria

COMMISSION OF THE EUROPEAN COMMUNITIES, The Identification of Irradiated Foodstuffs, Report of an International Colloquium, 1973, Eur 5126 (1974).

JOSEPHSON, E.S., "The use of ionizing radiation for preservation of food and feed products", Radiation Research, Proc. 5th Int. Congr. Seattle, 1975 (NYGAARD, O.S., ADLER, H.I., SINCLAIR, W.K., Eds), Academic Press, London (1975) 96–117.

BRYNJOLFSSON, A., "The national food irradiation program conducted by the US Department of the Army", Proc. Symp. Food Irradiation, 36th National Meeting of the Institute of Food Technologists, Anaheim, California, June 1976.

US ARMY NATICK RESEARCH AND DEVELOPMENT COMMAND, Radappertization (Radiation Sterilization) of Foods, Bibliography of Technical Publications and Papers, Technical Report, Natick TR-77/009 (1976).

Proc. EEC-Israel Workshop on the Application of Nuclear Techniques in Agriculture, Wageningen, 1976, Commission of the European Communities (1977).

WORLD HEALTH ORGANIZATION, Wholesomeness of Irradiated Food, Technical Report Series 604, WHO, Geneva (1977)
(44 pp., Sw.Fr. 6.00)

FOOD AND AGRICULTURE ORGANIZATION, Food Nutrition Series No.6, FAO, Rome (1977).

DIEHL, J.F., "Food irradiation", Radiation Processing (Trans. 1st Int. Mtg. Puerto Rico, 1976, SILVERMAN, J., VAN DYKEN, A., Eds), Radiat. Phys. Chem. 9 (1977) 193–206.

TOMITA, K., SUGIMOTO, S., "A commercial gamma ray irradiation plant in Japan", ibid. 567–73.

ELIAS, P.S., COHEN, A.J., Radiation Chemistry of Major Food Components, Elsevier, Amsterdam, Oxford, New York (1977).

FEDERATION OF AMERICAN SOCIETIES FOR EXPERIMENTAL BIOLOGY, Evaluation of the Health Aspects of Certain Compounds Found in Irradiated Beef, Bethesda, Maryland, USA (1977).

JOSEPHSON, E.S., PETERSON, M.S., Preservation of Food by Ionizing Radiation, Chemical Rubber Company Press, Cleveland (1978).

Food Irradiation in the United States, Report prepared for the Interdepartmental Committee on Radiation Preservation of Food, Washington, DC (1978).

Proc. National Symposium on Food Irradiation, 4–5 October 1979, Atomic Energy Board, Pretoria, South Africa.

Die Bestrahlung von Lebensmitteln und Futtermitteln (Vorträge der Informations-
tagung vom 23 September 1977 an der Eidgenössischen Technischen Hochschule in
Zürich), Chemische Rundschau, Sonderheft 1/78, Verlag Vogt-Schild AG, CH-4501
Solothurn 1, Switzerland.

JOINT FAO/WHO FOOD STANDARDS PROGRAMME, Recommended
International General Standard for Irradiated Foods and Recommended International
Code of Practice for the Operation of Radiation Facilities for the Treatment of
Foods, Codex Alimentarius Commission, CAC/RS 106-1979, CAC/RCP 19-1979,
FAO, Rome (1979).

PART II

LABORATORY EXERCISES

LABORATORY EXERCISES

INTRODUCTION

One of the best ways to become familiar with new techniques is to observe or actively participate in well organized laboratory demonstrations of the use of the technique. This is particularly true in the case of ionizing radiation as a food-processing method. The action of radiation on the food is quite different from that of heat-processing or other conventional methods. The colour, texture and flavour may be different, as the objective accomplished may be different from that gained by other commonly used processes. It is therefore considered highly desirable that certain types of demonstration be undertaken so that the trainee may observe the manner in which the treatment is carried out and the effects on the food product.

In some cases it may not be possible to allow the trainee to carry out the experiments independently of the instructor. For example, experiments involving microorganisms requiring very special techniques, and where maintaining aseptic conditions is important, cannot be taught to students within a single experiment. Therefore, only those with some training in microbiology could successfully carry out meaningful experiments on their own.

This part of the Manual contains a number of suggested exercises for the demonstration of the use of ionizing radiation in order to accomplish specific objectives in the treatment of food products for their preservation.

The greater part of these exercises constitute an essential part of the Training Courses on Food Irradiation organized by the International Facility for Food Irradiation Technology, Wageningen, Netherlands, annually since 1979. The exercises are intended to provide the trainees with laboratory experience in the following important areas of food irradiation: dosimetry, radiation chemistry, radiation microbiology, preservation of foods, sensory evaluation of foods, and control of insect infestation.

In planning these laboratory exercises it is necessary to have available the designated equipment and supplies. The requirements should be reviewed before undertaking the exercises.

DOSIMETRY PRACTICALS

In principle, any material showing a reliable, reproducible and measurable radiation effect can be used for radiation dosimetry. However, all systems generally accepted as reliable dose meters for absorbed dose measurements will have been subjected to a considerable amount of research, development and application in practical dosimetry. (For detailed information on using various dose-meter systems the reader is referred to Chapter V of the Manual of Food Irradiation Dosimetry, Technical Reports Series No. 178, IAEA, Vienna (1977)).

USE OF THE FRICKE FERROUS SULPHATE DOSE METER
TO DETERMINE DOSE RATE

PURPOSE

The aim of this experiment is to determine the dose rate in an unknown radiation field of a radionuclide source.

PRINCIPLE

The Fricke dose meter is the recommended reference dose-meter system for 'in-house' absorbed dose calibration in an irradiation facility.

The ferrous sulphate (Fricke) dose meter is based on the chemical process of oxidation of ferrous ions in acid aqueous solution to ferric ions by ionizing radiation. The method can be used for accurate absorbed dose determination from 40 to 400 Gy (4 to 40 krad), using spectrophotometric measurement of the ferric ion concentration at 305 nm wavelength, in the peak of the absorption spectrum. Since absorbance readings at a given wavelength may vary from one spectrophotometer to another, the molar extinction coefficient should be determined with each spectrophotometer set at the same wavelength and slit width used for the Fricke dosimetry measurements.

The change in absorbance (optical density) of an irradiated solution of ferrous ammonium sulphate in aqueous sulphuric acid is measured at this wavelength in a temperature-controlled spectrophotometer. Since the absorbed dose in units of grays or rads in the solution is linearly proportional to the change in absorbance, dose can be determined by multiplying the change in absorbance by a suitable conversion factor.

MATERIALS AND PREPARATION OF DOSIMETRY SOLUTION

1. Weigh 0.196 g of ferrous ammonium sulphate ($Fe(NH_4)_2(SO_4) \cdot 6H_2O$).
2. Weigh 0.03 g of sodium chloride.
3. Dissolve in 10 mltr of 0.8N H_2SO_4.
4. Dilute this to 500 mltr using 0.8N H_2SO_4 in a calibrated flask at 25°C.

(**Note:** This solution is not completely stable, but it can be stored in a clean, dark-brown, stoppered bottle at 15–20°C for up to eight weeks. A marked increase in the absorbance of the unirradiated solution at 305 nm wavelength indicates that the solution is no longer reliable.) The 0.8N H_2SO_4 is made by dissolving 22.5 mltr of concentrated sulphuric acid (density 1.84 $g \cdot cm^{-3}$) in distilled water to make 1 ltr of solution in a volumetric flask. (The acid is added little by little to the water — *never water to the acid.*) Glassware should be kept immaculately clean.

DOSE MEASUREMENT

1. Clean the capsules for irradiation by rinsing them with the dose-meter solution. Fill the capsules with the dose-meter solution and seal them.
2. Place the capsules carefully in the calibration holder in the batch irradiator.
3. Irradiate the capsules in the holder for a fixed time. Measure the time that the source is fully up very carefully. Note the time. (Make use of the Experimental Information Sheet.)

FRICKE – EXPERIMENTAL INFORMATION SHEET

Sample No.	Irrad./ not irrad.	Irrad. time	OD_{305}	t (°C)	Dose

4. Measure the optical density (OD) of the irradiated and unirradiated Fricke dose-meter solutions at a wavelength of 305 nm using the spectrophotometer. Note each value of OD. Check the temperature of each solution during each OD measurement. (A spectrophotometer with matched quartz cells (cuvettes) of 1 cm optical path-length, equipped with a temperature-controlled cell compartment, should be used for absorbance measurements.)
5. Calculate the dose given to each capsule using the equation:

$$D = \frac{N \cdot (OD_i - OD_0) \; 1.602 \times 10^{-17}}{\rho \, G \, \epsilon \, d \, (1 + 0.007 \, (t - 25))}$$

$$= \frac{2.75 \times 10^4 \, (OD_i - OD_0)}{1 + 0.007 \, (t - 25)} \quad \text{rad}$$

147

where N = 6.022 × 10^{23} (Avogadro's number)

ρ = 1.024 g·cm^{-3} (density of solution)

G = 15.6 × 10^{-2} eV^{-1} (the number of ferric ions formed by eV absorbed energy)

ϵ = 2195 ltr·mol^{-1}·cm^{-1} (OD of 1 mole/ltr of ferric ions in a 1-cm-long cuvette at 305 nm — it is the 'extinction' coefficient).

d = 1 cm (the length of the cuvette used for OD measurement)

t = temperature of the solution in degC at time of OD measurement

Laboratory Exercise 2

DETERMINATION OF DOSE RESPONSE OF A DOSE METER

PURPOSE

The aim of this experiment is to define the dose response curve of a practical dose-meter system over a useful dose range and to determine the reproducibility of the system.

PRINCIPLES

Almost all practical dose-meter systems are not absolute and must be calibrated against a standard system, usually in a position where the dose rate has been determined using a standard system. It is important to know whether or not the practical system has a linear calibration curve and how reproducible the system is. It is also important to know if the response to radiation is influenced by environmental factors.

MATERIALS

HX Perspex (PMMA) dose meters.
Spectrophotometer.
Linear graph paper.

DOSE-METER SYSTEM

The clear Perspex dose meter

Radiation induces a broad absorption band between 250 nm and 400 nm in clear polymethylmethacrylate (PMMA), often called Perspex, Lucite, Plexiglass. The dose meter is based on the measurement of the radiation-induced absorbance (optical density) at a suitable wavelength using a spectrophotometer. A wavelength of 314–315 nm is most suitable and it is best to use a specially made dosimetry Perspex.

To obtain reproducible results with clear PMMA dose meters, certain rules must be followed:

(a) The material should be carefully cleaned before use;

149

(b) Pre-irradiation background measurements should be made on the samples;

(c) Exposure of samples to light during or after irradiation should be avoided;

(d) Touching the optical faces of the dose meter should be avoided;

(e) Measurement should be made with a few hours of irradiation;

(f) Irradiation must take place under conditions of electronic equilibrium.

EXPERIMENTAL PROCEDURE

1. The PMMA samples (pieces of HX Perspex) must be cleaned with luke-warm dilute detergent solution, rinsed with clean water, dried on absorbent paper, numbered, and the optical density of the unirradiated (OD_0) PMMA dose meters measured at 314 nm using the spectrophotometer.
(**Note:** The optical density values of the unirradiated and irradiated samples are strongly dependent on wavelength.)

2. The PMMA samples of known optical density are irradiated under identical geometry as used with the Fricke ferrous sulphate system.

PERSPEX — EXPERIMENTAL INFORMATION SHEET

Sample No.	OD_0	Irrad. time	OD_i	Thickness	$OD_c = \Delta A_{corr}$

3. Note the irradiation time and measure the optical density at 314 nm of each of the samples (OD_i). Note these values on the Experimental Information Sheet. The induced change in absorbance, $\Delta A = OD_i - OD_0$, must be corrected to a value corresponding to standard thickness.

4. Measure the thickness (t) of each piece of Perspex at the position of the OD measurement using the micrometer. Note each thickness.

5. Calculate the thickness-corrected optical density (OD_c) for each of the samples using the equation:

$$OD_c = (OD_i - OD_0)\frac{1}{t} \qquad (t \text{ in mm})$$

Note: We are normalizing the OD_c to 1 mm thickness even though the samples are approximately 2 mm thick. The normalization to 1 mm is not essential; we

150

could choose 2 mm or 1 cm but it is wise to choose one value for normalization as different types of Perspexes have different thicknesses, e.g. 1 mm, 1.8 mm and 3 mm are common. It is important to note that calibration curves are often made for different normalization thicknesses and that different batches of Perspex can have slightly differing calibration curves even when they are normalized to the same thickness.

6. Calculate the mean value of OD_c $(\overline{OD_c})$ for the samples, which have been irradiated identically (at least 3 PMMA dose meters should be taken for the measurement of one and the same absorbed dose value).

7. Subsequently the corrected value of the induced change in absorbance, ΔA_{corr}, is read off the calibration curve for the corresponding wavelength (314 nm) and appropriate nominal thickness to obtain the absorbed dose value.

Note that the ΔA/dose relationship is not linear. It is therefore recommended that a calibration curve be made for each spectrophotometer used.

DOSE DISTRIBUTIONS IN THE IRRADIATED PRODUCT

PURPOSE

The aim of the exercise is to illustrate that in the irradiation of bulk material such as food an uneven distribution of dose always occurs through the product. Consequently the values and positions of D_{max} (maximum dose) and D_{min} (minimum dose) in the product in each type of irradiation facility should be known.

PRINCIPLE

The intensity of radiation decreases with distance from the source and with depth, by absorption, in the product. It is not possible to irradiate large volumes of bulk product with exactly the same dose everywhere in the product. Certain irradiation-facility techniques can be applied to improve the uniformity of the dose distribution such as movement of the product past the source in two dimensions, and two-sided irradiations (see Section 13.2). In the control of a food-irradiation process it is important to know the value and position of D_{max} and D_{min}.

MATERIALS

HX Perspex dose meters.
Spectrophotometer.
Calibration curve.

PROCEDURE

1. A series of numbered Perspex dose meters are attached on three stiff pieces of cardboard in the 25-point grid and placed in an irradiation product box of any appropriate dimensions so that two sheets form the outside planes and one sheet the central plane (see the block diagram, Fig. A).

2. The space between the pieces of cardboard is filled with a product (e.g. onions, rice, fruits or any other available food). Different foods may result in different bulk densities.

3. The Perspex dose meters are irradiated to measure the dose distribution in:

152

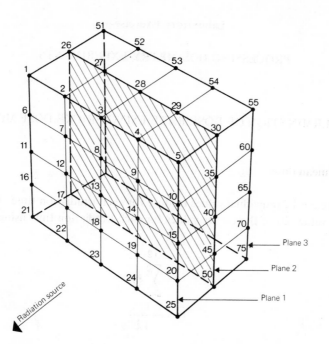

FIG. A. Dose distribution in an irradiated product (dose mapping). This product box has 75 distinct positions (to avoid confusion not all positions are put into this block diagram).

(a)   a product box which has been irradiated in a multipass irradiator system;
(b)   a product box which has been irradiated stationary and from only one side in an irradiator system.

4. The procedure for measuring the absorbed dose values of the dose meters is similar to that presented in Laboratory Exercise 2.

5. The absorbed dose values are plotted on the pieces of cardboard at their corresponding grids. The dose distribution in different lines and planes (depth and lateral doses) through the product boxes are observed in order to identify the positions of maximum dose (D_{max}) and minimum dose (D_{min}).

6. The uniformity ratio (U) is calculated using $U = D_{max}/D_{min}$.

Note: From this exercise it will be clear that, having determined the dose rate in a fixed position using the Fricke dose meter, and having set up a calibration curve for the Perspex using this position, and thus the doses given to the Perspex, you can irradiate the Perspex in an unknown radiation field and estimate the dose the Perspex has received.

153

Laboratory Exercise 4

PROCESSING DOSIMETRY EXPERIMENTS

1. DETERMINATION OF DOSE RATE USING FRICKE DOSE METER

1.1. The mean dose

There are 12 individual values of dose for each irradiation time. Determine the mean value ($\overline{D}$) of the twelve values for each irradiation time using

$$\overline{D} = \frac{\sum\limits_{i=1}^{12} D_i}{12}$$

1.2. The standard error of the mean

Determine the standard error (σ) of the mean dose for each irradiation time using

$$\sigma = \sqrt{\frac{\sum\limits_{i=1}^{12} (D_i - \overline{D})^2}{n(n-1)}} \quad \text{where } n = 12$$

1.3. Plot of dose versus time

Plot on linear graph paper the mean value of dose ($\overline{D}$) (vertical axis) against irradiation time (T) (horizontal axis). Draw in the standard error of the mean dose (σ) as an error in the dose value.

154

DATA SHEET FOR PART 1 OF LABORATORY EXERCISE 4

Irradiation time:
Mean dose $\bar{D}$ =

	Dose D_i	$D_i - \bar{D}$	$(D_i - \bar{D})^2$
1			
2			
3			
4			
5			
6			
7			
8			
9			
10			
11			
12			
$\sum D_i =$			$\sum$

$$\bar{D} = \frac{\sum D_i}{12} = \frac{}{12} = \qquad \text{rad}$$

$$\sigma = \sqrt{\frac{\sum (D_i - \bar{D})^2}{12.11}} = \sqrt{\frac{}{12.11}} =$$

1.4. The best straight line

Determine the best straight line through the four points. This is the equation

$$D = \dot{D}T + D_0$$

This will be done by linear regression using a minicomputer but you can calculate it yourself using:

155

$$\dot{D} = \frac{m \sum_{j=1}^{m} \overline{D}_m T_m - \sum_{j=1}^{m} \overline{D}_m \sum_{j=1}^{m} T_m}{m \sum_{j=1}^{m} T_m^2 - \sum_{j=1}^{m} T_m \sum_{j=1}^{m} T_m} = \text{dose rate}$$

where m = number of observations,

$\overline{D}_m$ = average of dose estimations,

T_m = irradiation time, and

$$D_0 = \frac{\sum_{j=1}^{m} \overline{D}_m \sum_{j=1}^{m} T_m^2 - \sum_{j=1}^{m} T_m \sum_{j=1}^{m} \overline{D}_m T_m}{m \sum_{j=1}^{m} T_m^2 - \sum_{j=1}^{m} T_m \sum_{j=1}^{m} T_m}$$

where m = 4 in this case.

(**Note**: the line can be even more accurately calculated by weighting each point with the reciprocal of the standard error of the mean.)

1.5. Draw the best straight line on your graph. The calibration point dose rate is   rad·h^{-1}. D_0 is   rad. What does D_0 represent?

2.   CALIBRATION CURVE FOR PERSPEX DOSE METER

2.1. Using the dose rate determined for the calibration position, estimate the total dose received by each set of Perspex samples from their irradiation time. Use the equation:

$$D = \dot{D}T + D_0$$

with $\dot{D} =$ and $D_0 =$

2.2. Plot on linear graph paper the mean value of corrected optical density ($\overline{OD_c}$) (vertical axis) against the relevant dose value (D) (horizontal axis).

156

DATA SHEET FOR PART 2 OF LABORATORY EXERCISE 4

Irrad. time, T	Dose, D	Mean OD, $\overline{OD}_c$

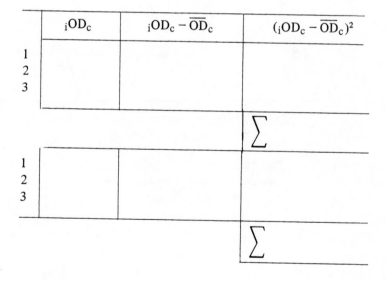

	$_iOD_c$	$_iOD_c - \overline{OD}_c$	$(_iOD_c - \overline{OD}_c)^2$
1			
2			
3			
			$\sum$
1			
2			
3			
			$\sum$

2.3. Determine the standard error of the mean of the corrected optical density for your own measurements using

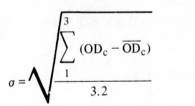

$$\sigma = \sqrt{\frac{\sum\limits_{1}^{3} (OD_c - \overline{OD}_c)}{3.2}}$$

2.4. Plot the standard error of the mean (σ) as an error in $\overline{OD}_c$ on the graph. If the percentage reproducibility of the Perspex dose meter is given by

$$\frac{\sigma}{\overline{OD}_c} \cdot 100\%$$

what is the percentage reproducibility in your measurement?

2.5. Is the calibration curve a straight line? Draw with the demonstrator the best smooth curve through the experimental points. This is the calibration curve.

Laboratory Exercise 5

ATTENUATION OF GAMMA RAYS (DISTANCE, ABSORBER)[1]

PURPOSE

The aim of the exercise is to reduce the intensity of γ-rays by increasing the distance from the radiation source and by bringing absorbing materials into the path of the γ-rays.

REQUIREMENTS

^{60}Co sources.
Lead absorbers.
Log-log graph paper (horizontal 3 and vertical 2 decades; Mercurius No. 7).
Semi-log graph paper (vertical 2 decades; Mercurius No. 4).

ATTENUATION BY DISTANCE

Introduction

For a point source, the radiation intensity is inversely proportional to the square of the distance if no intervening matter (solid, liquid or gaseous) is present in between. This is usually referred to as the inverse-square law. The intensity at a distance d from a point source (or other source with small dimensions in comparison with d) is thus given as

$$I_d^{\cdot} = k/d^2$$

where

$I_d^{\cdot}$ = intensity at d cm distance
k = proportionality constant (see Lecture Matter, Section 1.4.3)

In this exercise, diminution by distance of γ-rays from a 0.18 MBq (5 μCi) ^{60}Co source is investigated. The crystal scintillation counter is used.

---

[1] Experiment derived from Laboratory Training Manual on the Use of Isotopes and Radiation in Soil-Plant Relations Research, Laboratory Reports Series No. 29, IAEA, Vienna (1964) p. 70.

159

Procedure

1. Apply an operating voltage for the scintillation counter.
2. Determine the background count rate.
3. Determine the count rate as a function of distance for d = 20, 40,
60 ... 160 cm.
4. Plot the count rate resulting from sample (A* = R − r) against the
distance on log-log graph paper and draw the best straight line through the points.
5. Determine the slope of the line and explain the reason for discrepancies,
if any, by means of the inverse-square law.

ATTENUATION BY ABSORBERS

Introduction

The attenuation by matter of a collimated beam of monoenergetic γ-photons
is exponential. The attenuation of γ-rays from ^{60}Co which emits γ-photons of
slightly different energy (1.2 and 1.3 MeV) will be investigated.

Procedure

1. The previous experimental set-up is used except that the distance between
the sample and counter is now fixed.
2. Determine the count rate of the sample without adding any absorber
between source and counter.
3. Determine the count rate after placing one absorber between source and
counter and keep on repeating with increasing amounts of absorbers.
4. Remove the source completely and determine the background (with all
absorbers in place).
5. The net count rate from the sample is taken as a measure of the radiation
intensity, and this is plotted against linear absorber thickness on semi-log paper.
6. Determine the half-thickness of lead for ^{60}Co γ-rays and compare with
the table value.
7. Explain any discrepancy by means of simple exponential law (i.e. any
lack of straight line on semi-log paper).

PRESENTATION OF THE RESULTS

Note down on the appropriate table in the course laboratory:

1. Slope of the line in the attenuation by distance experiment.
2. Half-thickness of lead for ^{60}Co γ-rays in $g \cdot cm^{-2}$ and in cm (density of
lead: $11.34 \ g \cdot cm^{-3}$).

MICROBIOLOGY – DETERMINATION OF THE
D_{10} RADIATION DOSE OF *Escherichia coli*

PURPOSE

The aim of this exercise is to demonstrate the method of determining the sensitivity of microorganisms to radiation and to establish the dose necessary to reduce the population by a factor of 10, or one log cycle.

MATERIALS

Bacterial culture of *Escherichia coli* K 12 in phosphate buffer.
Peptone dilution fluid (0.1% peptone, 0.85% sodium chloride, pH 7.0).
Test tubes.
Petri plates.
Total plate count agar.
Incubator set at 37°C.
Water bath set at 46°C.

PROCEDURE

1. The instructor will prepare a suitable culture of the test organism in advance. This should have a population of 10^7 viable cells per mltr.

2. The organisms should be suspended in 1/15M phosphate buffer and from this the same amount placed in tubes appropriate for radiation treatment. The suspensions should be made in *duplicate* for each radiation dose selected, and for the control.

3. Irradiate with the following doses and then make decimal dilutions of the irradiated cultures using peptone dilution fluid:

Dose (Gy)	Dilutions to prepare		
0	10^{-4}	10^{-5}	10^{-6}
100	10^{-3}	10^{-4}	10^{-5}
200	10^{-2}	10^{-3}	10^{-4}
300	10^{-2}	10^{-3}	10^{-4}
400	10^{-1}	10^{-2}	10^{-3}
500	10^{-0}	10^{-1}	10^{-2}

4. Pipette 1 mltr of each dilution in duplicate into Petri plates and pour at once with the total plate count agar. The agar must have been held in the 46°C water bath to ensure a uniform temperature, so that all organisms will receive the same heat shock at the time of pouring of the plates.

5. When the agar has solidified, invert the plates and incubate them for two days at 37°C.

RESULTS

1. Count all plates having between 30 and 300 colonies. Average the counts of the duplicate plates. Note the data (use the Information Sheet).

2. Plot the log of the surviving organisms against the radiation dose received.

3. From the curves constructed from the results of the experiment, determine the D_{10} value by reading the number of Gy necessary to reduce the initial population by one log cycle.

If one plots the $\log_{10}$ of the number of survivors versus the dose (semilog plot) one gets a straight line in the exponential portion of the 'survivor curve'. D_{10} is the reciprocal of the slope of the straight line portion, or the dose for the curve to traverse one logarithmic cycle (the 90% destruction or 10% survival dose). Calculate from the logarithmic values of the straight line portion the regression line by the least-squares method. This regression line $y = a + bx$ gives the D_{10} value ($D = -1/b$). Note D_{10}, r and the number of test points (use the Information Sheet).

QUESTIONS

1. Give the important factors in characterizing dose requirements for microbial destruction.

2. The cells were irradiated in the presence of air. By which factor does the D_{10} value increase if the cells had been irradiated under anoxic conditions?

3. The cells of *E. coli* were collected in the stationary phase. Do you expect a higher or lower radiation resistance of log-phase cells?

4. Give some radiation survival curves representing a non-exponential rate of death. Explain the 'shoulder' and the 'tail'.

REFERENCE

See Lecture Matter of this Manual, Section 5.2.2.

LABORATORY EXERCISE 6, EXPERIMENTAL INFORMATION SHEET
Radiation resistance of *E. coli* K 12 in phosphate buffer (1/15M, pH 7) under aerobic conditions

Dilution	Radiation treatment (Gy)					
	0	100	200	300	400	500
10^{-0}						
10^{-1}						
10^{-1}						
10^{-2}						
10^{-2}						
10^{-3}						
10^{-3}						
10^{-4}						
10^{-4}						
10^{-5}						
10^{-5}						
10^{-6}						
10^{-6}						

D_{10} value (Gy)	Regression coefficient (r)	No. of test points

MICROBIOLOGY – DETERMINATION OF THE D_{10} RADIATION DOSE OF SPORES OF *Bacillus stearothermophilus*

PURPOSE

The purpose of this exercise is to demonstrate the radiation resistance of bacterial spores.

MATERIALS

Spore-suspension of *B. stearothermophilus.*
Peptone dilution fluid (0.1% peptone, 0.85% sodium chloride, pH 7.0).
Test tubes.
Petri plates.
Dextrose tryptone agar.
Incubator, set at 55°C.
Water bath set at 45°C.

PROCEDURE

1. A suspension of *B. stearothermophilus* NCA 1518 spores in $(1/15)$ mol·ltr^{-1} phosphate buffer (pH 7.0) will be provided at a concentration of approximately 1×10^7 per mltr.

2. Heat-shock this spore-suspension for 20 minutes at 80°C, cool and then place 2-mltr portions into sterile tubes. Each treatment in duplicate.

3. Irradiate with the following doses and then make decimal dilutions of the irradiated cultures using peptone dilution fluid:

Dose (kGy)	Dilutions to prepare			
0	10^{-5}	10^{-6}		
2	10^{-3}	10^{-4}	10^{-5}	10^{-6}
4	10^{-2}	10^{-3}	10^{-4}	10^{-5}
6	10^{-1}	10^{-2}	10^{-3}	10^{-4}
8	10^{-0}	10^{-1}	10^{-2}	10^{-3}
10	10^{-0}	10^{-1}	10^{-2}	

4. Pipette 1 mltr of each dilution in duplicate into Petri plates and pour at once with dextrose tryptone agar. The agar must have been held in the 45°C water bath.

5. When the agar has solidified, invert the plates, put them in a plastic bag (to prevent drying-up) and place them in the 55°C incubator. Incubate for four days.

LABORATORY EXERCISE 7, EXPERIMENTAL INFORMATION SHEET
Radiation resistance of *B. stearothermophilus* in phosphate buffer $((1/15)$ mol/ltr, pH 7) under aerobic conditions

Dilution	Radiation treatment (kGy)					
	0	2	4	6	8	10
10^{-1}						
10^{-1}						
10^{-2}						
10^{-2}						
10^{-3}						
10^{-3}						
10^{-4}						
10^{-4}						
10^{-5}						
10^{-5}						
10^{-6}						
10^{-6}						

Decimal reduction dose D_{10} =   Gy

   =   krad

RESULTS

Determine the D_{10} value according to the method described in Laboratory Exercise 6. Note data on Information Sheet. Plot the log of the number of survivors versus the dose for getting the survivor curve.

QUESTIONS

1. The bacterial spores were irradiated in the presence of air. By which factor does the D_{10} value increase if the spores had been irradiated under anaerobic conditions?

2. Do you expect a correlation between the heat and radiation resistance of microorganisms?

3. The dose requirement for radappertization is determined by the micro-organism associated with the food that has the greatest radiation resistance. Give the name of this organism for non-acid low-salt foods. Give the radiation dose needed for sterilization.

REFERENCE

GOULD, G.W., HURST, A., The Bacterial Spore, Academic Press, London (1969).
See Lecture Matter of this Manual, Section 5.2.2.

MICROBIOLOGY – THE EFFECT OF IRRADIATION ON THE MICROBIOLOGICAL QUALITY OF MINCED MEAT

PURPOSE

The aim of this exercise is to demonstrate the reduction in the number of pathogenic and spoilage microorganisms in a food caused by a radiation dose of 2.5 kGy (250 krad).

PRINCIPLE

Ionizing radiation is lethal to bacteria and is generally due to the action of the hydroxyl radical, OH, produced from radiolysis of water. The doses needed to produce a given degree of killing depend chemically on, among other factors, the particular microorganism and the nature of the food itself.

MATERIALS

Minced meat (50% pork, 50% beef).
Colworth Stomacher 400 blender.
Sterile plastic bags.
Petri dishes.
Pipettes.
Water bath set at 46°C.
Incubators at 17, 25, 30 and 37°C.
Peptone dilution fluid (0.1% peptone, 0.85% NaCl).
Sterile glass spatula.

Enumeration of bacterial species	Media (Oxoid)	Trade Nos.
Total aerobic count	Plate count agar	(PC agar) (CM 325)
Gram-negative rod-shaped microorganisms	Plate count agar plus crystal violet solution (2 mg·ltr^{-1})	(PCV agar) (CM325)
Enterobacteriaceae	Violet red bile glucose agar	(VRBG agar) (CM 485)
E. coli	MacConkey agar	(MC agar) (CM 115)

Moulds and yeasts	Oxytetracycline glucose yeast-extract agar	(OGYE agar) (CM 545)
Coagulase-positive staphylococci	Baird-Parker medium	(BP agar)* (CM 275)
Faecal streptococci	Azide blood agar base plus crystal violet solution (2 mg·ltr^{-1})	(AB agar)* (CM 259)

* BP agar and AB agar are prepared one day before use and poured into plates.

PROCEDURE FOR PREPARATION OF A HOMOGENATE OF IRRADIATED AND NON-IRRADIATED SAMPLE

Weigh aseptically 10 g of minced meat into a sterile plastic bag. Add 90 mltr dilution fluid, blend for one minute with the Stomacher 400. This is the 10^{-1} dilution.

Wait five minutes for sedimentation and make the following further dilution range using peptone dilution fluid:

Dose	Dilution to make
Control	10^{-2} 10^{-3} 10^{-4} 10^{-5} 10^{-6}
2.5 kGy	10^{-2} 10^{-3} 10^{-4}

PROCEDURE FOR ENUMERATION OF TOTAL AEROBIC MESOPHILIC VIABLE COUNT

1. Pipette 1 mltr from the dilutions 10^{-3}, 10^{-4}, 10^{-5}, 10^{-6} of the control and 10^{-1}, 10^{-2}, 10^{-3}, 10^{-4} of the irradiated sample in duplicate into the Petri dishes and pour with plate count agar.

2. When the agar has solidified, invert the plates and place in the 30°C incubator for three days.

3. Count the plates which contain 30 to 300 colonies and multiply the count by the appropriate dilution factor to get the total plate count per gram minced meat. Note the data on the Information Sheet.

PROCEDURE FOR ENUMERATION OF GRAM-NEGATIVE ROD-SHAPED MICROORGANISMS

Use the same procedure as for total aerobic mesophilic viable count, except that these plates have to be incubated at 17°C for five days. Note results on the Information Sheet.

PROCEDURE FOR ENUMERATION OF ENTEROBACTERIACEAE

1. Pipette 1 mltr from the dilutions 10^{-2}, 10^{-3}, 10^{-4}, 10^{-5} of the control and the dilutions 10^{-1}, 10^{-2}, 10^{-3} of the irradiated sample in duplicate in Petri plates and pour at once with violet red bile glucose agar; mix and, after solidification, pour a second layer with the same medium.

2. Incubate the plates upside down at 30°C for 48 hours.

3. Calculate the number of viable enterobacteriaceae per gram minced meat. Colonies are considered to be positive for enterobacteriaceae if they are surrounded by a zone of precipitation accompanied by a purple discoloration of the layer. Note the data on the Information Sheet.

PROCEDURE FOR ENUMERATION OF E. coli

1. Pipette 1 mltr from the dilutions 10^{-1}, 10^{-2}, 10^{-3} of the control and the dilutions 10^{-1} of the irradiated sample in duplicate in Petri plates and pour at once with MacConkey agar and mix.

2. Incubate the plates upside down at 44°C for 24 hours.

3. Calculate the number of E. coli (intense violet-red colonies).

PROCEDURE FOR ENUMERATION OF MOULDS AND YEASTS

1. Pipette 1 mltr from the dilutions 10^{-1}, 10^{-2}, 10^{-3} of the control and 10^{-1}, 10^{-2} of the irradiated sample in duplicate in Petri plates and pour at once with oxytetracycline yeast-extract glucose agar.

2. Incubate the plates upside down at 25°C for three to five days.

3. Calculate the number of yeasts and moulds (examples of plates with yeasts and moulds will be shown by the instructor). Note the data on the Information Sheet.

PROCEDURE FOR ENUMERATION OF COAGULASE-POSITIVE STAPHYLOCOCCI

1. Pipette 0.1 mltr from the dilution 10^{-1}, 10^{-2}, 10^{-3} of the control and 10^{-1}, 10^{-2} of the irradiated sample in duplicate on the surface of the prepared Petri dishes containing Baird-Parker medium and spread it, using a sterile glass spatula.

2. Incubate the plates upside down at 37°C for 24 hours.

LABORATORY EXERCISE 8, EXPERIMENTAL INFORMATION SHEET
Microbiological quality of irradiated and non-irradiated minced meat

Dilution	Total aerobic count		Gram-negative rods		Entero-bacteriaceae		E. coli		Moulds		Yeasts		Staphylo-cocci		Faecal streptococci	
	Control	2.5 kGy	Control	2.5 kGy	Control	2.5 kGy	Control	2.5 kGy	Control	2.5 kGy	Control	2.5 kGy	Control	2.5 kGy	Control	2.5 kGy
10^{-1}																
10^{-1}																
10^{-2}																
10^{-2}																
10^{-3}																
10^{-3}																
10^{-4}																
10^{-4}																
10^{-5}																
10^{-5}																
10^{-6}																
10^{-6}																
Count per gram minced meat																

TABLE A. MICROBIOLOGICAL LIMITS FOR GOOD QUALITY OF MINCED MEAT (from Mossel and Tamminga, 1973)

Aerobic-plate count	10^6
Gram-negative rods	10^5
Enterobacteriaceae	10^3
E. coli	10
Salmonellae	1/20 gram
S. aureus	10^3
Faecal streptococci	10^3
Sulphite-reducing clostridia	10^3
C. perfringens	10^2

3. Calculate the number of staphylococci; only black colonies are considered to be staphylococci if they have a clear zone and within this zone an opaque zone. Note data on the Information Sheet.

PROCEDURE FOR ENUMERATION OF FAECAL STREPTOCOCCI

1. Pipette 0.1 mltr from the dilution 10^{-1}, 10^{-2}, 10^{-3}, 10^{-4} of the control and 10^{-1}, 10^{-2} of the irradiated sample in duplicate on the surface of the prepared Petri dishes containing Azide blood agar base and spread it, using a sterile glass spatula.

2. Incubate the plates upside down at 37°C for three to five days.

3. Calculate the number of faecal streptococci; colonies are considered to be faecal streptococci if they are small, smooth, round and mostly violet. Note data on the Information Sheet.

QUESTIONS

1. Explain the purpose of this microbiological testing scheme and compare the results with the microbiological limits for good quality (see Table A).

2. Is it necessary to resuscitate the cells from irradiated food before inoculation into or onto selective media?

3. Give the confirmation tests for enterobacteriaceae, E. coli, Staphylococcus aureus and faecal streptococci.

4. For which purpose can the total enterobacteriaceae be used as index group?

5. Give the two main objectives of a low-dose application to fresh meat.

REFERENCES

The Oxoid Manual, 4th edn, Oxoid (1979).

MOSSEL, D.A.A., TAMMINGA, S.K., Methoden voor het Microbiologisch Onderzoek van Levensmiddelen, Noordervliet, Zeist (1973).

INTERNATIONAL ATOMIC ENERGY AGENCY, Microbiological Specifications and Testing Methods for Irradiated Food, Technical Reports Series No.104, IAEA, Vienna (1970).

SPECK, M.L., (Ed.) Compendium of Methods for the Microbiological Examination of Foods, American Public Health Association, Washington (1976).

MOSSEL, D.A.A., Microbiology of Foods and Dairy Products, University of Utrecht (1975).

See Lecture Matter of this Manual, Section 7.3.1.

EVALUATION OF EXTENSION OF MARKET LIFE
OF FRESH FISH BY IRRADIATION

PURPOSE

This exercise is intended to illustrate (a) the effect of radiation treatment on the extension of the market life of a food normally held at refrigeration temperatures, and (b) the changes in microbial flora that bring about food deterioration.

MATERIALS

40 fillets (or steaks) of non-fatty fish, not to exceed 500 g each.
20 clear plastic film wraps or bags.
20 clear plastic film bags, suitable for vacuum packing.
Refrigerator, set at 3°C.
120 sterile Petri plates for bacterial counts.
120 sterile saline dilution blanks (99 mltr after sterilization).
120 sterile 1-mltr pipettes.
Sterile total plate count agar (standard methods).
Incubator, set at 30°C.

PROCEDURE

1. Package 20 fillets in plastic bags or wraps (air-pack).
2. Package 20 fillets in plastic bags and seal under vacuum.
3. Place five coded packages of air packs and five coded packages of vacuum packs in refrigerator.
4. Immediately irradiate five coded packages of air packs and five coded packages of vacuum packs at a radiation dose of 1.5 kGy (150 krad). (It is best to code the packages for the week of intended withdrawal, as well as the treatment.)
5. Place all packages in refrigerator.

EVALUATION

1. As soon as practicable, remove the four zero-storage-time samples from the refrigerator for evaluation.
2. Aseptically open each package and remove an appropriate sample for microbiological analysis. Proceed as under (4) below.

3. Quickly assess the quality factors indicated in the suggested evaluation chart, using a scale for each factor selected by the instructor. To gain experience, these evaluations should be made and recorded by each individual.

4. Make a microbiological analysis of the sample previously taken under (2) above.

    (a) Aseptically weigh 10 g of sample into a sterile electric blender jar containing 90 mltr of sterile saline dilution water.

    (b) Blend just long enough to obtain a homogeneous sample, but not long enough for the sample to heat up.

    (c) Make appropriate dilutions by placing a measured amount of the homogenate into sterile dilution water, to obtain three decimal dilutions to give countable plates of bacterial colonies. The instructor should have sufficient experience to make a selection of the dilutions required.

    (d) Transfer the dilutions to sterile Petri plates and at once pour the total plate count agar into the plates and mix thoroughly.

    (e) As soon as the agar has solidified, invert the plates and place in the incubator at 30°C. (The instructor should explain why each step is taken, so the student may be fully aware of the reasons for the whole procedure, since the student will not be able to perform the microbiological examination for himself.)

    (f) After three days of incubation, visually examine the plates to see if they are ready for counting. If the plates are not being overgrown, wait and count the plates after five days of incubation.

    (g) Count plates having between 30 and 300 colonies and calculate the number present in each sample by multiplying the count by the dilution factor. Record the count for each sample.

RESULTS

1. Repeat the examination of samples every week, using the same methods used for the zero-storage-time samples (see above).

2. After the first week of storage, the dilution used for the microbial examination of the irradiated and non-irradiated samples will have to be adjusted to take into account the changes in the microbial populations of the samples.

3. Plot the results of the various factors studied in the entire experiment against the time of storage.

4. From these curves determine the maximum extension of market life afforded by the packaging and the radiation treatment.

REFERENCE

See Lecture Matter of this Manual, Section 7.3.2.

Laboratory Exercise 10

EFFECT OF IRRADIATION ON SPROUTING OF POTATOES

PURPOSE

The aim of this exercise is the evaluation of the effect of irradiation on the sprouting ability of potatoes.

PRINCIPLE

A low dose of radiation can inhibit sprouting of potatoes and onions. This effect can be made visible in one week by using the accelerating effect of gibberellic acid on the sprouting by breaking the dormancy.

MATERIALS

Early harvested potatoes.
Special potato soil.
Gibberellic acid A3, aqueous solution of 1 mg$\cdot$ltr^{-1}.
Incubation room at 20°C.
'Pommes parisiennes' knife.

PROCEDURE

1. Irradiate a part of the potatoes with 0.1 kGy (10 krad).
2. From each potato cut out some eyes, using the special knife.
3. Place the pieces of potatoes on a tray and let dry for two hours at room temperature.
4. Place the dried potato pieces in the gibberellic acid solution for ten minutes.
5. Dry the pieces at room temperature for 16 hours.
6. Plant the potato pieces in the soil (1 to 2 cm deep).
7. Incubate for seven days in the dark at 20°C.
8. Score the number of sprouts.

QUESTIONS

1. What are the advantages and disadvantages of using radiation for sprout inhibition of potatoes?

175

2. What is the average maximum storage time of potatoes either untreated or treated by irradiation and chemical sprout inhibitors respectively?

3. What could be the toxicological problems arising from the use of chemicals in sprout inhibition?

REFERENCES

LEWIS, N.F., MATHUR, P.B., Extension of storage lives of potatoes and onions by cobalt-60 rays, Int. J. Appl. Radiat. Isotopes **14** (1963) 443.

See Lecture Matter of this Manual, Section 7.3.4.

NAIR, P.M., et al., "Studies of sprout inhibition of onions and potatoes and delayed ripening of bananas and mangoes by gamma irradiation", Radiation Preservation of Food (Proc. Symp. Bombay, 1972), IAEA, Vienna (1973) 347.

BRUINSMA, SINNEMA, BAKKER, ZWART, Europ. Potato J. **10** (1967) 136.

RAPPAPORT, LIPPERT, TRIM, Am. Potato J. **34** (1957) 244.

Laboratory Exercise 11

EXTENSION OF MARKET LIFE OF STRAWBERRIES BY IRRADIATION

PURPOSE

The aim of this exercise is to demonstrate control of the growth of certain moulds by irradiation treatment and use of commercial packaging.

PRINCIPLE

Irradiation can inactivate moulds, resulting in a retardation of spoilage and an increase of shelf life. A closed packaging prevents desiccation.

MATERIALS

Absolutely fresh strawberries (approx. 15 kg) that show no evidence of spoilage.

Commercial boxes and chipboard boxes with a cover provided with a cellulose acetate window.

Storage rooms set for 10°C (selling temperature) and 20°C (room temperature); relative humidity (RH) about 90%.

Dose meters, PMMA Perspex (see Laboratory Exercise 2).

Spectrophotometer.

PROCEDURE

1. Repack the product in commercial boxes (200 g) and chipboard boxes (250 g).
2. Code the boxes.
3. Provide some boxes with dose meters.
4. Irradiate and store the boxes according to the following scheme:

Dose	Commercial boxes		Chipboard boxes	
	10°C	20°C	10°C	20°C
0 Gy (0 rad)	5	5	5	5
1.5 kGy (150 krad)	5	5	5	5
2.5 kGy (250 krad)	5	5	5	5

5. After irradiation treatment, place the boxes in the storage rooms at indicated temperatures.

6. Set aside one box of each of the above treatments for taste-testing.

RESULTS

1. Measure and calculate the absorbed dose (see Laboratory Exercise 2).

2. Examine all boxes for evidence of moulding or spoilage at various time intervals during storage. Do not disturb the berries in the boxes. Record results.

3. When about 20% of the berries are affected by mould, pour the berries out of the boxes and examine them individually.

4. Record number of berries showing: (a) marked moulding, (b) trace of moulding and (c) no moulding.

5. Calculate the percentage of berries in these three categories.

6. Determine the treatment giving the best results and estimate the possible extensions of market life.

7. Prepare coded berries from the boxes mentioned in Point 6 of the Procedure for testing by the trainees one day after irradiation and ask them to evaluate the flavour score of each berry. Record the combined scores for each treatment.

QUESTIONS

1. At which stage of ripening will an irradiation treatment give the best results?

2. Which factors will appoint the maximum dose to control mould in strawberries?

3. In what way will a closed packaging improve the quality of irradiated strawberries?

REFERENCES

CHALUTZ, E., MAXIE, F.C., SOMMER, N.F., The interaction of gamma irradiation, sealed packages, and controlled atmosphere on incidence of *Botrytis* rot in strawberries, USAEC Research and Development Rep. UCD-34 P 80-3, 9-26 (1965).

BERAHA, L., et al., Gamma radiation in the control of decay of strawberries, grapes and apples, Food Technol. **15** (1961) 94.

Laboratory Exercise 12

EXTENSION OF MARKET LIFE OF MUSHROOMS BY IRRADIATION

PURPOSE

The aim of this exercise is to demonstrate the effect of irradiation and packaging on the keeping qualities of mushrooms.

PRINCIPLE

Irradiation delays senescence and prevents the caps of mushrooms from opening. The preservative effect of irradiation is assisted by appropriate packaging in lessening desiccation.

MATERIALS

Absolutely fresh mushrooms showing no open caps.
Commercial boxes and chipboard boxes with cover provided with a cellulose-acetate window.
Storage rooms set for 10°C (selling temperature) and 20°C (room temperature); relative humidity (RH) about 90%.
Dose meters, PMMA Perspex (see Laboratory Exercise 2).
Spectrophotometer.

PROCEDURE

1. Repack the product in commercial boxes (200 g) and chipboard boxes (250 g).
2. Code the boxes.
3. Provide some boxes with dose meters.
4. Irradiate and store the boxes according to the scheme shown below:

Dose	Commercial boxes		Chipboard boxes	
	10°C	20°C	10°C	20°C
0 Gy (0 rad)	4	4	4	4
1.5 kGy (150 krad)	4	4	4	4
2.5 kGy (250 krad)	4	4	4	4

5. After irradiation treatment, place the boxes in the storage rooms at indicated temperatures.

RESULTS

1. Measure and calculate the absorbed dose (see Laboratory Exercise 2).
2. Examine all boxes for quality deterioration (discoloration, stalks growing, caps opening, gills darkening) at various time intervals during storage.
3. When about 20% of the mushrooms show open caps pour the mushrooms out of the boxes and examine them individually.
4. Record number of mushrooms showing: (a) open caps and (b) no open caps.
5. Calculate the percentage of mushrooms in these two categories.
6. Determine the treatment giving the best results and estimate the possible extension of shelf-life.

QUESTIONS

1. What is the effect of irradiation treatment on the quality of mushrooms?
2. What is the advantage of a sealed package in combination with irradiation?
3. Does a retardation of the senescence run parallel to the sensory qualities of the irradiated mushrooms? What about the non-irradiated mushrooms?

REFERENCES

CAMPBELL, J.D., STOTHERS, S., VAISEY, M., BERCK, B., Gamma irradiation influence on the storage and nutritional quality of mushrooms, J. Food Sci. **33** (1968) 540.

GILL, W.J., NICHOLAS, R.C., MARKAKIS, P., Irradiation of cultured mushrooms, Food Technol. **23** (1969) 111.

KOVÁCS, E., VAS, K., FARKAS, J., Attempts to extend useful storage life of champignon by ionizing radiations, Kiserl. Közl. Elelmiszeripar 1−3, 3−17 (1968).

MARKAKIS, P., Irradiation of mushrooms, Mushroom News **17** (1969) 1.

LANGERAK, D.I., "The influence of irradiation and packaging upon the keeping quality of fresh mushrooms", Proc. 8th Int. Congress on Mushroom Science (1971) 221.

EXTENSION OF MARKET LIFE OF TOMATOES
BY A COMBINATION OF HEAT AND IRRADIATION

PURPOSE

The aim of this exercise is to demonstrate the effect of heat, irradiation, and both in combination, on the inactivation of moulds occurring on fruit and vegetables.

PRINCIPLE

The preservative effect of irradiation can often be advantageously combined with the effect of heat. The resulting combination treatment may involve synergistic or cumulative action of the combination partners, leading to a decrease in the treatment required for one or both agents.

MATERIALS

Sound, freshly harvested tomatoes of one of the locally available varieties, and with degrees of ripeness usually identified by the colour stages 'turning' and 'pink'.

Container with tap-water.

Scalpels, stopwatch.

Spore suspension of *Botrytis cinerea:* concentration approx. 10^5 spores/mltr.

Micropipettes of 0.005 mltr.

Waterbaths set at 20°C, 40°C and 45°C.

Incubator set at 25°C.

Trays.

Polyethylene bags.

Storage room set at 20°C and relative humidity (RH) 90%.

PROCEDURE

1. Remove the corolla of the fruit.
2. Wash the tomatoes during 5 seconds in tap-water and dry them on filter paper.

3. On two sides of the tomatoes make an incision 1/2 cm long and 2 to 3 mm deep.
4. Code the samples per treatment.
5. Inoculate the incisions with 0.005 mltr of the spore suspension (approx. 500 spores) and incubate the tomatoes for 24 hours at 25°C.
6. Apply the heat, irradiation or combined treatment according to the following scheme:

Treatment		Colour stage	
Waterbath	Irradiation doses (kGy)	Turning	Pink
No dipping	0, 0.5, 1	3 × 10	3 × 10
5 s at 20°C	0, 0.5, 1	3 × 10	3 × 10
5 s at 40°C	0, 0.5, 1	3 × 10	3 × 10
5 s at 45°C	0, 0.5, 1	3 × 10	3 × 10

Total number of selected tomatoes is 240.

7. Place the tomatoes on trays (one treatment per tray) and store them at 20°C and 90% RH.

RESULTS

1. Examine all tomatoes for evidence of mould at various time intervals during storage.
2. Record number of tomatoes showing: (a) mould, (b) no mould.
3. Calculate the percentage of tomatoes in these two categories per treatment.
4. Determine the treatment giving the best results and estimate the possible extensions of shelf-life.

QUESTIONS

1. What factors are important with regard to the optimal effect of an irradiation treatment controlling moulds on tomatoes?
2. Explain the phenomenon that in irradiated unripe tomatoes the percentage mould rot can be higher than in non-irradiated tomatoes.

REFERENCES

BEN-ARIE, R., BARKAI-GOLAN, R., Combined heat-radiation treatments to control storage rot of Spadona pears, Int. J. Appl. Radiat. Isotopes **20** (1969) 687.

LANGERAK, D.I., CAÑET PRADES, F.M., The effect of combined treatment on the inactivation of moulds in fruit and vegetables, Tech. and Prelim. Research Report No. 88, Foundation ITAL, Netherlands (1979).

Lecture Matter of this Manual, Section 7.3.3.

Laboratory Exercise 14

DETERMINATION OF EFFECTS OF RADIATION ON ASCORBIC ACID (VITAMIN C) IN FOOD

PURPOSE

The aim of this exercise is to demonstrate the effect of ionizing radiation on vitamin C with respect to the nature of the substrate in which the compound is contained, the total dose absorbed and the direct and indirect actions involved (see Section 4.3.5 of Lecture Matter).

MATERIALS

Fresh orange juice.
Glacial HPO_3 pellets.
Glacial acetic acid.
Crystalline ascorbic acid (VSP reference ascorbic acid). Keep cool, dry and cut off direct sunlight.
2,6 dichloroindophenol sodium salt.
Soda lime.
$NaHCO_3$.
Filter paper, plain and fluted.
Amber glass-stoppered bottles, 500 mltr.
3 Erlenmeyer flasks, 50 mltr.
Burette, 50 mltr.
Volumetric flasks and pipettes.
Desiccator.
Distilled water.

PROCEDURE

This is the method for ascorbic acid of the Association of Official Agricultural Chemists and is applicable to food products provided that they do not contain ferrous, stannous or cuprous ions, SO_2 or thiosulphate:

Preparation of reagents

(1) Metaphosphoric acid — acetic acid stabilizing extracting solution. Dissolve, with shaking, 15 g of glacial HPO_3 pellets, or freshly pulverized stick

HPO_3, in 40 mltr of acetic acid and 200 mltr distilled water. Dilute to 500 mltr and filter rapidly through fluted filter paper into a glass-stoppered bottle. (HPO_3 slowly changes to H_3PO_4 but if stored in refrigerator will remain satisfactory for seven to ten days.)

(2) Indophenol standard solution. Dissolve 50 mg of 2,6 dichloroindophenol sodium salt that has been stored in a desiccator over soda lime in 50 mltr of distilled water to which has been added 42 mg of $NaHCO_3$. Shake vigorously, and when the dye is dissolved, dilute to 200 mltr. Filter through fluted filter paper into an amber glass-stoppered bottle. Keep the bottle, stoppered and out of direct light, in a refrigerator.

Note: Check this solution by adding 5 mltr of the extracting solution, (1) above, containing ascorbic acid, to 15 mltr of the indophenol dye reagent. If the reduced solution is not practically colourless, discard and prepare a fresh stock solution. If the dry dye is at fault, use a new batch.

(3) Reference standard ascorbic acid solution. Weigh accurately (±0.1 mg) about 100 mg of crystalline ascorbic acid, transfer to a 100-mltr glass-stoppered volumetric flask and dilute to volume with HPO_3-HOAc reagent at room temperature.

(4) Standardize the indophenol solution at once as follows: Transfer three 2-mltr aliquots of the ascorbic acid solution to each of three 50-mltr Erlenmeyer flasks containing 5 mltr of the HPO_3-HOAc reagent. Titrate rapidly with the indophenol solution from the 50-mltr burette until a light but distinct rose-pink colour persists for at least five seconds. (Each titration should require about 15 mltr of solution.) Titration should check within 0.1 mltr.

Similarly titrate three blanks composed of 7 mltr of the HPO_3-HOAc reagent plus a volume of the H_2O equivalent to the volume of indophenol solution used in the direct titration.

After subtracting the average of the blanks (usually about 0.1 mltr) from the standardization titrations, calculate and express the concentration of indophenol solution as 1 mg ascorbic acid equivalent to 1 mltr of reagent. The indophenol solution should be standardized daily with a freshly prepared solution of reference standard ascorbic acid solution.

Determination of ascorbic acid in the orange juice

Express juice from a number of oranges to yield at least one litre of juice. Mix thoroughly and set aside half of the juice in an amber bottle in the refrigerator for irradiation. Filter the other half through absorbant cotton or rapid paper filter, and analyse as follows:

Add aliquots of at least 100 mltr of the orange juice to an equal volume of the HPO_3-HOAc reagent. Mix and filter rapidly through rapid folded paper (Eaton-Dikeman No.195, 18.5 cm, or equivalent). Titrate 10-mltr aliquots

DATA SHEET FOR ASCORBIC ACID IRRADIATION (LAB. EXERCISE 14)

Dose (Gy)	Orange juice		Distilled water ascorbic acid solution		Distilled water solution of irradiated dry ascorbic acid crystals	
	mg ascorbic acid per 100 mltr	Destruction (%)	mg ascorbic acid per 100 mltr	Destruction (%)	mg ascorbic acid per 100 mltr	Destruction (%)

with the standardized indophenol solution and make blank determinations for correction of the titrations, as described above, using appropriate volumes of HPO_3-HOAc reagent and H_2O. Express ascorbic acid concentration as mg/100 mltr of original sample.

Determination of ascorbic acid in a known solution

(1) Weigh out an amount of crystalline ascorbic acid in 500 mltr of distilled water, to approximate the ascorbic acid content found in the orange juice above.

(2) Add 100-mltr aliquots of the known ascorbic acid solution to 100 mltr of the HPO_3-HOAc reagent. Titrate 10-mltr aliquots with the standardized indophenol solution, make blank determinations, and express the results as mg/100 mltr of ascorbic acid.

Demonstration of the effect of radiation on ascorbic acid

(1) Dispense three 100-mltr aliquots of the original orange juice that was set aside for irradiation and four 100-mltr aliquots of the distilled water solution of ascorbic acid into thoroughly cleaned glass containers suitable for irradiation (e.g. 25 X 200-mm borosilicate glass culture tubes). Place some ascorbic acid crystals in three dry tubes such as were used for the liquids. Each tube should contain the same weighed amount of crystalline ascorbic acid, to make a 500-mltr solution of approximately the same ascorbic acid concentration as the orange juice and the water solution.

(2) Set aside one of the tubes of the water solution of ascorbic acid to act as a non-irradiated control.

(3) Irradiate one of the tubes of each of the three lots (juice, water solution and dry crystals) at each of the radiation doses of 100 Gy, 1 kGy and 10 kGy (10 krad, 100 krad and 1 Mrad).

(4) After the irradiation treatment, carefully transfer the weighed amount of crystalline ascorbic acid to 500-mltr volumetric flasks and fill to volume with distilled water.

(5) Add the 100 mltr of each of the nine irradiated samples to 100 mltr of the HPO_3-HOAc reagent. Titrate 10-mltr aliquots of each with the standardized indophenol solution and calculate the ascorbic acid content of each sample.

(6) Make the same titration with the sample of the water solution of ascorbic acid that was set aside as the non-irradiated control. Record results in Data Sheet.

RESULTS

Evaluation of the effect of radiation on ascorbic acid under different conditions

From the results obtained by the various titrations of ascorbic acid in the different solutions:

(a) Compare the effect of the radiation dose in each of the types of solutions of ascorbic acid;

(b) Compare the effect of radiation at the same dose for each of the different types of solutions of ascorbic acid.

It should be found that the loss of ascorbic acid in each type of solution is roughly proportional to the absorbed radiation dose.

Plot ascorbic acid concentration against absorbed dose.

The effect of a given dose on the three different lots of ascorbic acid should be in the following order of increasing destruction:

Dry crystals < whole orange juice < distilled water solution

This illustrates the direct effect in the crystals and the indirect effect due to free radicals in the solutions.

Laboratory Exercise 15

EFFECT OF GAMMA RADIATION ON INACTIVATION
OR KILLING OF INSECTS

PURPOSE

The aim of this exercise is to determine the radiosensitivity of insects at different stages of development.

PRINCIPLE

Larval, pupal or adult stages differ as to radiosensitivity and the survival scored after six days will indicate this. The difference in radiosensitivity between Dipteran and Coleopteran species will also be shown.

MATERIALS

Insect species used:
Onion fly (pupae and flies); *Hylemya antiqua* Meigen.
Bean beetle (larvae, pupae and adults); *Acanthoscelides obtectus* Say.
Grain weevil (adults); *Sitophilus granaria.*
Tribolium castaneum (adults).

EQUIPMENT (for 16 students):

Three stereomicroscopes, CO_2 cylinder, Petri dishes, insect cages, five trays to store the irradiated samples, tweezers, razor blades, scalpels, three counters, beans, corn meal, labels, brushes, insect-sucking tubes.

PROCEDURE

1. Place a fixed number, but at least 25, adults or certain life stages of the adult of the species mentioned above into Petri dishes (3 replicates). (Note that adult onion flies have to be anaesthetized by CO_2 in the cages before being put in the Petri dishes. Use sucking tube.)
2. Place a fixed number, but at least 25, beans provided by the instructor into Petri dishes (3 replicates per item).

3. Irradiate (γ-rays) the samples with 0, 100, 250, 500, 1000 and 1500 Gy (2 Petri dishes of each species, life stages or beans per irradiation dose).

4. After irradiation, put the adult onion flies into the cages again with food and water.

5. All Petri dishes and cages are stored at room temperature for six days.

RESULTS

After six days the survival rate is analysed.

1. Count dead and almost dead organisms as well as the number of insects (adult stage) that emerged from the beans or pupae. Where necessary the stereomicroscope is used (up to 25× magnification).

2. Investigate larvae in the irradiated and non-irradiated beans by using razor blade, scalpel, tweezers and stereomicroscope, and determine their viability.

Note: experience should be gained from a separate sample of control beans before starting with the beans of this exercise.

3. Interpret the results obtained from the various irradiation treatments using the correction of Abbott:

$$\text{Corrected mortality} = \frac{X - Y}{X} \cdot 100$$

X = percentage *alive* in the control

Y = percentage *surviving* in treated samples

REFERENCES

Lecture Matter of this Manual, Section 5.2.4.

SCHEDULE: LAB. EXERCISE 15

	0 Gy	100 Gy (10 krad)	250 Gy (25 krad)	500 Gy (50 krad)	1000 Gy (100 krad)	1500 Gy (150 krad)
Onion fly (pupae)	x	x	x	x		
Onion fly (young adults)	x	x		x	x	
Bean beetle (young adults)	x	x		x	x	
Bean beetle (larvae/pupae) in beans, before emergence	x	x	x	x	x	
Grain weevil (adult)	x			x		x
Tribolium (young adult)	x				x	x

Laboratory Exercise 16

TASTE PANEL EXPERIMENTS

(A) Threshold value test

PURPOSE

The aim of this test (single-stimulus method) is to determine the average detection level of a group of panel members for the four primary tastes.

PRINCIPLE

Sixteen bottles contain solutions of several taste substances in different concentrations. The panel member should detect whether these solutions are sweet (sugar), salty (NaCl), bitter (caffein) or sour (tartaric acid). Some of the concentrations are so high that nearly everybody can recognize them; others are very weak.

MATERIALS

Sixteen solutions of the four taste substances, each in four concentrations, are prepared in coded bottles from 1 to 16. The concentrations $(mol \cdot ltr^{-1})$ used are the following:

Sugar	Salt	Tartaric acid	Caffein
0.004	0.004	0.00005	0.0002
0.012	0.012	0.00040	0.0008
0.036	0.036	0.00320	0.0032
0.108	0.108	0.02560	0.0128

Bottles of tap-water.
Beaker glasses.
Plastic beakers.
Buckets.
Threshold value test sheets for scoring results (one for each panel member), as shown.

191

THRESHOLD VALUE TEST SHEET: LABORATORY EXERCISE 16

Bottle No.	Sweet	Salty	Bitter	Sour	Questionable	Water
1						
2						
3						
4						
5						
6						
7						
8						
9						
10						
11						
12						
13						
14						
15						
16						

PROCEDURE

1. A bottle containing a certain taste solution and a bottle containing tap-water stand before you.

2. Rinse your mouth (use the glass beaker) with tap-water.

3. Pour about 40 mltr of the taste solution from the bottle into a plastic beaker.

4. Taste it thoroughly by swirling it around in your mouth. Don't swallow it, but spit it out into a bucket beside you.

5. Note the perceived taste on the sheet by writing a cross in the correct square (i.e. corresponding bottle line and corresponding taste column). If you hesitate, mark the solution in the column under 'questionable'. If you cannot detect any taste, mark the solution as 'water'.

6. Wait for the instructor's sign, and hand the bottle with the taste solution to your right-hand neighbour.

7. Rinse your mouth again with tap-water.

8. Pour 40 mltr from the new bottle into a new beaker.

9. Taste it and note its taste again on your test sheet.
10. Pass the bottle to your right.
11. Proceed in this way until all bottles have passed you.

EVALUATION OF RESULTS

The results of the individual panel members, scored on the threshold value test sheets, are combined and expressed in a suitable table and graph.

(B) Triangular (difference) test

PURPOSE

The aim of this test is to determine whether or not a difference exists between the organoleptic properties of two samples.

PRINCIPLE

Assessors (panel members) are asked to select the odd sample out of three coded ones, two of which are equal and one of which may be different. The selection should be based on general organoleptic properties, e.g. odour, flavour or appearance.

MATERIALS

Irradiated or otherwise treated food product and an untreated sample.
Beakers.
Tap-water and beaker for rinsing.
Buckets.
Tissues.
Test sheets.

PROCEDURE

1. Three samples stand before you.
2. Taste as accurately as possible.
3. Select a different sample.

TABLE B. TRIANGULAR (DIFFERENCE) TEST: PROBABILITY TABLE

Number of tasters	Number of correct answers necessary to establish the significance of a result as indicated by: *	**	***	Number of tasters	Number of correct answers necessary to establish the significance of a result as indicated by: *	**	***
1	—	—	—	26	14	15	17
2	—	—	—	27	14	16	18
3	3	—	—	28	15	16	18
4	4	—	—	29	15	17	19
5	4	5	—	30	15	17	19
6	5	6	—	31	16	18	20
7	5	6	7	32	16	18	20
8	6	7	8	33	17	18	21
9	6	7	8	34	17	19	21
10	7	8	9	35	17	19	22
11	7	8	10	36	18	20	22
12	8	9	10	37	18	20	22
13	8	9	11	38	19	21	23
14	9	10	11	39	19	21	23
15	9	10	12	40	19	21	24
16	9	11	12	41	20	22	24
17	10	11	13	42	20	22	25
18	10	12	13	43	21	23	25
19	11	12	14	44	21	23	25
20	11	13	14	45	22	24	26
21	12	13	15	46	22	24	26
22	12	14	15	47	23	24	27
23	12	14	16	48	23	25	27
24	13	15	16	49	23	25	28
25	13	15	17	50	24	26	28

* Significant difference (see text).
** Highly significant difference (see text).
*** Very highly significant difference (see text).

(Data from BENGTSSON, K., Modern methods of sensory analysis, Wallerstein Lab. Commun. **16** (1953) 231—250.)

194

4. Note the result on the test sheet.
5. Rinse your mouth.

EVALUATION OF RESULTS

The chance of achieving adequate sorting in triangular tests depends on the extent of the differences between the samples. If the differences are considerable, then nobody will have difficulty in sorting, but when the differences are small, tasters are no longer capable of sorting. In an effort to give an answer nevertheless, they begin to gamble. The chance of guessing right is 0.33 in this test. On the basis of theoretical considerations, which cannot be discussed here, a Probability Table (Table B) has been established for interpretation of results of triangular tests. This table shows the minimum numbers of correct sorting, for a given number of tasters, necessary to establish significant differentation at three significance levels (at three levels of probability of error, P):

* means $0.01 < P < 0.05$ (the corresponding differences between samples are known as 'significant');
** means $0.001 < P < 0.01$ (the corresponding differences between samples are known as 'highly significant');
*** means $P < 0.001$ (the corresponding differences between samples are known as 'very highly significant').

If, for example, from sixteen tasters nine or more appeared to select the odd sample correctly, the differences between the samples are significantly present. If lower, there is not sufficient certainty to warrant the decision that samples are different.

GLOSSARY

absorbed dose, D. The absorbed dose (sometimes referred to simply as dose), D, is the amount of energy absorbed per unit mass of irradiated matter at a point in the region of interest. More formally, it may be defined as the quotient of $d\epsilon$ by dm, where $d\epsilon$ is the mean energy imparted by ionizing radiation to matter of mass dm. The special name for SI unit of absorbed dose is gray (Gy); the rad may be used temporarily.

absorbed dose rate, $\dot{D}$. The increment of absorbed dose in a particular medium during a given time interval.

aqueous electron, e_{aq}^-. The hydrated electron, a product of the radiolysis of water.

batch irradiator. A non-continuous irradiator in which the irradiation process must be stopped, the treated product removed, and the next batch of untreated product inserted for irradiation. See **stationary irradiator.**

becquerel (Bq). The SI derived unit of activity, being one radioactive disintegration per second of time. It has dimensions of s^{-1}, and its relationship to the traditional special unit, the curie (Ci), is:

$$1 \text{ Bq} = 2.7\dot{0}2\dot{7} \times 10^{-11} \text{ Ci}$$

The term disintegration refers to a nuclear transformation, i.e. either a change of nuclide or an isomeric transition.

bulk density. Weight per unit volume of the product 'en masse', as it would be irradiated (e.g. with potatoes; the air space between the potatoes is also considered in determining the bulk density).

bulk irradiator. Irradiator with facilities (transport system) for handling produce in bulk (e.g. potatoes, grain, onions, etc.)

calibration curve. The response curve (radiation effect as a function of absorbed dose) established under controlled conditions in which the doses are determined by comparison with a standard reference dose meter.

197

Codex Alimentarius. A collection of internationally adopted food standards (codes of practice, guidelines, etc.) resulting from the activities of the Joint FAO/WHO Food Standards Programme as implemented by the Codex Alimentarius Commission, on which 117 states were represented in 1979.

curie (Ci). The special unit of activity, which is being superseded by the becquerel (Bq). The curie is defined as:

$$1 \text{ Ci} = 3.7 \times 10^{10} \text{ disintegrations per second}$$
$$= 3.7 \times 10^{10} \text{ Bq}$$

The term disintegration refers to a nuclear transformation, i.e. either a change of nuclide or an isomeric transition.

$\overline{D}_{min}$, $\overline{D}_{max}$. Mean minimum and maximum absorbed doses in the product. These are determined from the frequency distributions (normally Gaussian) of a number of determinations of the minimum and maximum absorbed doses, D_{min} and D_{max} respectively, measured in the food product.

disinfestation. Control of the proliferation of insect and other pests in grain, cereal products, dried fruit, spices, etc. This requires a dose of $\sim$ 0.3 to 1 kGy ($\sim$ 30 to 100 krad), which will usually kill pests in all life-cycle stages.

dose, absorbed, *see* **absorbed dose.**

dose distribution. The spatial variation in absorbed dose throughout the product, the dose having the extreme values D_{max} and D_{min}.

dose meter. A device, instrument or system having a reproducible and measurable response to radiation that can be used to measure or evaluate the quantity termed absorbed dose, exposure or similar radiation quantity. [The word dosimeter has been replaced by dose meter as standard terminology (dose meter, but dosimetry, dosimetric)].

dose uniformity, U (*or* **dose uniformity ratio**). The ratio of maximum to minimum absorbed dose in the product, i.e. $U = D_{max}/D_{min}$.

dosimetry. The measurement of radiation quantities, specifically absorbed dose and absorbed dose rate, etc.

dwell time. The time the product remains in each irradiation position in a shuffle-dwell irradiator.

electron accelerator. A device for imparting large amounts of kinetic energy to electrons.

electron beam. An essentially monodirectional stream of (negative) electrons which have usually been accelerated electrically or electromagnetically to high energy.

electronvolt (eV). A unit of energy. One electronvolt is the kinetic energy acquired by an electron in passing through a potential difference of one volt in a vacuum. It is defined as:

$$1 \text{ eV} = 1.60219 \times 10^{-19} \text{ J, approximately}$$

It is a unit used with the SI whose value is obtained experimentally. A commonly used multiple is mega-electronvolt (10^6 eV = 1 MeV).

FAO. Food and Agriculture Organization of the United Nations, Rome.

film badge. Small radiographic film in light-tight envelope worn by personnel working in radiation areas to register exposure to ionizing radiation.

free radical. An electrically neutral atom or molecule with an unpaired electron in the outer orbit. A dot placed in a manner as shown in OH· (the OH radical) designates a free radical.

geometry, *see* **irradiation geometry**.

gray (Gy). The SI derived unit of absorbed dose of ionizing radiation, being equal to one joule of energy absorbed per kilogram of matter undergoing irradiation. It has dimensions of $J \cdot kg^{-1}$, and its relationship to the traditional special unit, the rad, is:

$$1 \text{ Gy} = 1 \text{ J} \cdot kg^{-1} = 100 \text{ rad}$$

G-value. The radiation yield of chemical changes in an irradiated substance in terms of the number of specified chemical changes produced per 100 eV or per joule of energy absorbed from ionizing radiation. Examples of such chemical changes are production of particular molecules, free radicals, ions, etc. In the case of molecules affected, it is sometimes called molecular yield.

At present, most data are given as number of chemical changes per 100 eV. To convert to SI units, transform to per-eV value and divide by 1.602×10^{-19} to obtain data in per-joule value, e.g.

$$15.6 \text{ per } 100 \text{ eV} \rightarrow 15.6 \times 10^{-2} \text{ eV}^{-1} \rightarrow \left(\frac{15.6 \times 10^{-2}}{1.602 \times 10^{-19}} \right) \text{ J}^{-1} \rightarrow 9.74 \times 10^{17} \text{ J}^{-1}$$

high dose. In food irradiation, doses of 10 kGy or more.

IAEA. International Atomic Energy Agency, Vienna.

ionization. Production of ion pairs, one of which may be an electron.

ionizing radiation. Any radiation, consisting of directly or indirectly ionizing particles, or a mixture of both.

irradiation geometry. The spatial description of the relative positions of the product and the radiation source during the radiation treatment (comprises source-to-target distance, sizes, spacing, shape, position of scattering or shielding material, etc.).

irradiator. That part of the radiation facility that houses the radiation source and associated equipment, i.e. the radiation chamber inside the radiation protection shield.

isotopes. Nuclides having the same atomic number (i.e. the same chemical element) but having different mass number (i.e. same Z, different A).

labyrinth. A passage linking two areas that is designed to follow a tortuous path such that no radiation originating in one area can pass into the other area without undergoing at least a single reflection or absorption at a passage/wall interface.

low dose. In food irradiation, doses up to about 1 kGy.

medium dose. In food irradiation, doses of about 1 to 10 kGy.

nuclide. Any given atomic species characterized by: (1) the number of protons, Z, in the nucleus; (2) the number of neutrons, N, in the nucleus; and (3) the energy stage of the nucleus (in the case of an isomer).

rad (rad). The special unit of absorbed dose, which is being superseded by the SI unit of dose, the gray (Gy). The rad is defined as:

$$1 \text{ rad} = 0.01 \text{ Gy} = 0.01 \text{ J} \cdot \text{kg}^{-1} \ (= 100 \text{ erg} \cdot \text{g}^{-1})$$

radappertization. The application to foods of doses of ionizing radiation sufficient to reduce the number and/or activity of viable microorganisms to such an extent that very few, if any, are detectable in the treated food by any recognized method (viruses being excepted). In the absence of post-processing contamination, no microbial spoilage or toxicity should become detectable with presently available methods, no matter how long or under what conditions the food is stored. Dose ranges used are ~ 10 to 50 kGy (~ 1 to 5 Mrad).

radiation. To be understood as referring to ionizing radiation in this Manual.

radiation energy. The spectral energy of the particles in the radiation beam. The beam may be monoenergetic, comprise particles of a number of discrete energies, or comprise a mixture of energies giving rise to a continuous energy spectrum.

radiation process. As applied to food irradiation, the act of irradiating a product in order to treat it in a beneficial way (e.g. to improve its intrinsic or commercial value, or to extend its keeping qualities, etc.).

radiation source. An apparatus or radioactive substance in a suitable support that constitutes the origin of the ionizing radiation (e.g. cobalt-60 source rods in a frame, or an electron accelerator).

radicidation. The practical elimination of pathogenic organisms and microorganisms other than viruses by means of irradiation. It is achieved: (i) by the destruction of organisms like tape-worm and trichina in meat, for which doses range between 0.1 and 1 kGy (10 and 100 krad); and (ii) by the reduction of the number of viable specific non-spore-forming pathogenic microorganisms, such that none is detectable in the treated food by any standard method for which doses range between 2 and 8 kGy (200 and 800 krad).

radionuclide. A radioactive nuclide.

radurization. The application to foods of doses of ionizing radiation sufficient to enhance keeping quality (usually at refrigeration temperature) by causing a substantial decrease in numbers of viable specific spoilage microorganisms. Dose ranges used are ~ 0.4 to 10 kGy (~ 40 krad to 1 Mrad).

rem (rem). The special unit of dose equivalent used for radiation protection purposes only. It is being superseded by the SI unit of dose equivalent, the sievert (Sv). Its dimensions are joules per kilogram (i.e. energy per mass).

roentgen (R). The special unit of exposure, defined as:

$$1 \text{ R} = 2.58 \times 10^{-4} \text{ C} \cdot \text{kg}^{-1}$$

There is no SI derived unit for this quantity.

sievert (Sv). The SI unit of dose equivalent used for radiation protection purposes only. It has dimensions of joules per kilogram, and its relationship to the traditional special unit, the rem, is:

$$1 \text{ Sv} = 100 \text{ rem}$$

source strength. Of a gamma-ray radiation source, the strength defines the activity of the radioactive nuclide source material; it is expressed in becquerels (or curies).

stationary irradiator. Irradiation facility in which neither the source nor the product moves during the period the product is undergoing irradiation, i.e. there is no automatic product transport system (e.g. conveyor); a batch type of operation, where the handling of the product is carried out 'manually', using fork-lift trucks, trolleys, etc. The source has to be shut off during such operations, i.e. in the case of radionuclide sources these have to be moved into their shielded store.

two-sided irradiation. Irradiation of a product from two opposite sides.

uniformity ratio, *see* **dose uniformity**.

utilization efficiency. The fraction of radiation energy emitted that is absorbed by the total product after the completion of an irradiation cycle.

WHO. World Health Organization, Geneva.

wholesomeness. In connection with irradiated foods, their safety for human consumption must be based on the following considerations: (a) the absence of microorganisms and microbial toxins harmful to man; (b) the nutritional contribution to the total diet of the irradiated food; (c) the absence of any significant amounts of toxic products formed in the food as a result of the irradiation process.

The following conversion table is provided for the convenience of readers

FACTORS FOR CONVERTING SOME OF THE MORE COMMON UNITS TO INTERNATIONAL SYSTEM OF UNITS (SI) EQUIVALENTS

NOTES:

(1) SI base units are the metre (m), kilogram (kg), second (s), ampere (A), kelvin (K), candela (cd) and mole (mol).

(2) ▶ indicates SI derived units and those accepted for use with SI;
▷ indicates additional units accepted for use with SI for a limited time.
[*For further information see the current edition of The International System of Units (SI), published in English by HMSO, London, and National Bureau of Standards, Washington, DC, and International Standards ISO-1000 and the several parts of ISO-31, published by ISO, Geneva.*]

(3) The correct symbol for the unit in column 1 is given in column 2.

(4) ✳ indicates conversion factors given exactly; other factors are given rounded, mostly to 4 significant figures:
≡ indicates a definition of an SI derived unit: [ ] in columns 3+4 enclose factors given for the sake of completeness.

Column 1 Multiply data given in:	Column 2	Column 3 by:	Column 4 to obtain data in:	
Radiation units				
▶ becquerel	1 Bq	(has dimensions of s^{-1})		
disintegrations per second (= dis/s)	1 s^{-1}	$\equiv 1.00 \times 10^0$	Bq	✳
▷ curie	1 Ci	$= 3.70 \times 10^{10}$	Bq	✳
▷ roentgen	1 R	$[= 2.58 \times 10^{-4}$	C/kg]	✳
▶ gray	1 Gy	$[\equiv 1.00 \times 10^0$	J/kg]	✳
▷ rad	1 rad	$= 1.00 \times 10^{-2}$	Gy	✳
▶ sievert *(radiation protection only)*	1 Sv	$[\triangleq 1.00 \times 10^0$	J/kg]	✳
▷ rem *(radiation protection only)*	1 rem	$= 1.00 \times 10^{-2}$	Sv	✳
Mass				
▶ unified atomic mass unit ($\frac{1}{12}$ of the mass of ^{12}C)	1 u	$[= 1.660\,57 \times 10^{-27}$	kg, approx.]	
▶ tonne (= metric ton)	1 t	$[= 1.00 \times 10^3$	kg]	✳
pound mass (avoirdupois)	1 lbm	$= 4.536 \times 10^{-1}$	kg	
ounce mass (avoirdupois)	1 ozm	$= 2.835 \times 10^1$	g	
ton (long) (= 2240 lbm)	1 ton	$= 1.016 \times 10^3$	kg	
ton (short) (= 2000 lbm)	1 short ton	$= 9.072 \times 10^2$	kg	
Length				
statute mile	1 mile	$= 1.609 \times 10^0$	km	
▷ nautical mile (international)	1 n mile	$= 1.852 \times 10^0$	km	✳
yard	1 yd	$= 9.144 \times 10^{-1}$	m	✳
foot	1 ft	$= 3.048 \times 10^{-1}$	m	✳
inch	1 in	$= 2.54 \times 10^1$	mm	✳
mil (= 10^{-3} in)	1 mil	$= 2.54 \times 10^{-2}$	mm	✳
Area				
▷ hectare	1 ha	$[= 1.00 \times 10^4$	m^2]	✳
▷ barn *(effective cross-section, nuclear physics)*	1 b	$[= 1.00 \times 10^{-28}$	m^2]	✳
square mile, (statute mile)^2	1 mile^2	$= 2.590 \times 10^0$	km^2	
acre	1 acre	$= 4.047 \times 10^3$	m^2	
square yard	1 yd^2	$= 8.361 \times 10^{-1}$	m^2	
square foot	1 ft^2	$= 9.290 \times 10^{-2}$	m^2	
square inch	1 in^2	$= 6.452 \times 10^2$	mm^2	
Volume				
▶ litre	1 l *or* 1 L	$[= 1.00 \times 10^{-3}$	m^3]	✳
cubic yard	1 yd^3	$= 7.646 \times 10^{-1}$	m^3	
cubic foot	1 ft^3	$= 2.832 \times 10^{-2}$	m^3	
cubic inch	1 in^3	$= 1.639 \times 10^4$	mm^3	
gallon (imperial)	1 gal (UK)	$= 4.546 \times 10^{-3}$	m^3	
gallon (US liquid)	1 gal (US)	$= 3.785 \times 10^{-3}$	m^3	

This table has been prepared by E.R.A. Beck for use by the Division of Publications of the IAEA. While every effort has been made to ensure accuracy, the Agency cannot be held responsible for errors arising from the use of this table.

Column 1 Multiply data given in:	Column 2	Column 3 by:	Column 4 to obtain data in:
Velocity, acceleration			
foot per second (= fps)	1 ft/s	$= 3.048 \times 10^{-1}$	m/s ✱
foot per minute	1 ft/min	$= 5.08 \ \times 10^{-3}$	m/s ✱
mile per hour (= mph)	1 mile/h	$=\begin{cases}4.470 \times 10^{-1} \\ 1.609 \times 10^{0}\end{cases}$	m/s km/h
▷ knot (international)	1 knot	$= 1.852 \times 10^{0}$	km/h ✱
free fall, standard, g		$= 9.807 \times 10^{0}$	m/s^2
foot per second squared	1 ft/s^2	$= 3.048 \times 10^{-1}$	m/s^2 ✱
Density, volumetric rate			
pound mass per cubic inch	1 lbm/in^3	$= 2.768 \times 10^{4}$	kg/m^3
pound mass per cubic foot	1 lbm/ft^3	$= 1.602 \times 10^{1}$	kg/m^3
cubic feet per second	1 ft^3/s	$= 2.832 \times 10^{-2}$	m^3/s
cubic feet per minute	1 ft^3/min	$= 4.719 \times 10^{-4}$	m^3/s
Force			
▶ newton	1 N	$[\equiv 1.00 \ \times 10^{0}$	m·kg·s^{-2}] ✱
dyne	1 dyn	$= 1.00 \ \times 10^{-5}$	N ✱
kilogram force (= kilopond (kp))	1 kgf	$= 9.807 \times 10^{0}$	N
poundal	1 pdl	$= 1.383 \times 10^{-1}$	N
pound force (avoirdupois)	1 lbf	$= 4.448 \times 10^{0}$	N
ounce force (avoirdupois)	1 ozf	$= 2.780 \times 10^{-1}$	N
Pressure, stress			
▶ pascal^a	1 Pa	$[\equiv 1.00 \ \times 10^{0}$	N/m^2] ✱
atmosphere^b, standard	1 atm	$= 1.013\ 25 \times 10^{5}$	Pa ✱
▷ bar	1 bar	$= 1.00 \ \times 10^{5}$	Pa ✱
centimetres of mercury (0°C)	1 cmHg	$= 1.333 \times 10^{3}$	Pa
dyne per square centimetre	1 dyn/cm^2	$= 1.00 \ \times 10^{-1}$	Pa ✱
feet of water (4°C)	1 ftH$_2$O	$= 2.989 \times 10^{3}$	Pa
inches of mercury (0°C)	1 inHg	$= 3.386 \times 10^{3}$	Pa
inches of water (4°C)	1 inH$_2$O	$= 2.491 \times 10^{2}$	Pa
kilogram force per square centimetre	1 kgf/cm^2	$= 9.807 \times 10^{4}$	Pa
pound force per square foot	1 lbf/ft^2	$= 4.788 \times 10^{1}$	Pa
pound force per square inch (= psi)^c	1 lbf/in^2	$= 6.895 \times 10^{3}$	Pa
torr (0°C) (= mmHg)	1 torr	$= 1.333 \times 10^{2}$	Pa
Energy, work, quantity of heat			
▶ joule ($\equiv$ W·s)	1 J	$[\equiv 1.00 \ \times 10^{0}$	N·m] ✱
▶ electronvolt	1 eV	$[= 1.602\ 19 \times 10^{-19}$	J, approx.]
British thermal unit (International Table)	1 Btu	$= 1.055 \times 10^{3}$	J
calorie (thermochemical)	1 cal	$= 4.184 \times 10^{0}$	J ✱
calorie (International Table)	1 cal$_{IT}$	$= 4.187 \times 10^{0}$	J
erg	1 erg	$= 1.00 \ \times 10^{-7}$	J ✱
foot-pound force	1 ft·lbf	$= 1.356 \times 10^{0}$	J
kilowatt-hour	1 kW·h	$= 3.60 \ \times 10^{6}$	J ✱
kiloton explosive yield (PNE) ($\equiv 10^{12}$ g-cal)	1 kt yield	$\simeq 4.2 \ \ \times 10^{12}$	J

---

^{a} Pa (g): pascals gauge ^{b} atm (g) (= atü): atmospheres gauge ^{c} lbf/in^2 (g) (= psig): gauge pressure
Pa abs: pascals absolute   atm abs (= ata): atmospheres absolute   lbf/in^2 abs (= psia): absolute pressure

Column 1 *Multiply data given in:*	Column 2	Column 3 *by:*	Column 4 *to obtain data in:*

Power, radiant flux

▶ watt | 1 W | $[\equiv 1.00 \times 10^0$ | $J/s]$ ✳
British thermal unit (International Table) per second | 1 Btu/s | $= 1.055 \times 10^3$ | W
calorie (International Table) per second | 1 cal$_{IT}$/s | $= 4.187 \times 10^0$ | W
foot-pound force/second | 1 ft·lbf/s | $= 1.356 \times 10^0$ | W
horsepower (electric) | 1 hp | $= 7.46 \times 10^2$ | W   ✳
horsepower (metric) (= ps) | 1 ps | $= 7.355 \times 10^2$ | W
horsepower (550 ft·lbf/s) | 1 hp | $= 7.457 \times 10^2$ | W

Temperature

▶ kelvin | K
▶ degrees Celsius, t | $t = T - T_0$ | | ✳
    where T is the thermodynamic temperature in kelvin and T_0 is defined as 273.15 K

degree Fahrenheit | $t_{°F} - 32$ | | t *(in degrees Celsius)* ✳
degree Rankine | $T_{°R}$ | $\left. \right\} \times \left(\dfrac{5}{9}\right)$ gives | T *(in kelvin)* ✳
temperature difference[d] | $\Delta T_{°R}\,(= \Delta t_{°F})$ | | $\Delta T\ (= \Delta t)$ ✳

Thermal conductivity[d]

1 Btu·in/(ft²·s·°F) | *(International Table Btu)* | $= 5.192 \times 10^2$ | $W·m^{-1}·K^{-1}$
1 Btu/(ft·s·°F) | *(International Table Btu)* | $= 6.231 \times 10^3$ | $W·m^{-1}·K^{-1}$
1 cal$_{IT}$/(cm·s·°C) | | $= 4.187 \times 10^2$ | $W·m^{-1}·K^{-1}$

Miscellaneous quantities

litre per mole per centimetre | (1 M/cm =) 1 L·mol^{-1}·cm^{-1} | $= 1.00 \times 10^{-1}$ m²/mol   ✳
(molar extinction coefficient or molar absorption coefficient)
G-value, traditionally quoted per 100 eV
  of energy absorbed | 1×10^{-2} eV^{-1} | $= 6.24 \times 10^{16}$ | J^{-1}
(radiation yield of a chemical substance)
mass per unit area | 1 g/cm² | $[= 1.00 \times 10^1$ | $kg/m^2\,]$ ✳
(absorber thickness and mean mass range)

---

[d] A temperature interval or a Celsius temperature difference can be expressed in degrees Celsius as well as in kelvins.